# CURSE ME ONCE, WINKLE ME TWICE

A Kerilee Oberon Paranormal Cozy Mystery

Petra Shaw

**CSD Digital Enterprises, LLC**

Copyright © 2025 Petra Shaw

All rights reserved

The characters and events portrayed in this book are fictitious. Any similarity to real persons, living or dead, is coincidental and not intended by the author.

No part of this book may be reproduced, or stored in a retrieval system, or transmitted in any form or by any means, electronic, mechanical, photocopying, recording, or otherwise, without express written permission of the publisher.

ISBN-13: 9781234567890
ISBN-10: 1477123456

Cover design by: Art Painter
Library of Congress Control Number: 2018675309
Printed in the United States of America

# CONTENTS

Title Page

Copyright

Chapter 1: Windmill with the Sunflower Door — 1

Chapter 2: He Possesses the Bodies of Animals — 18

Chapter 3: The Diner Chronicles — 26

Chapter 4: Rip and the Blue Demon — 33

Chapter 5: The Empty Jug — 43

Chapter 6: Same Diner, Different Day — 56

Chapter 7: The Mahtantu Effect! — 68

Chapter 8: Adrienna's Discovery — 79

Chapter 9: He Who Tarries — 92

Chapter 10: Roger's Dream — 108

Chapter 11: Cataloging Garden Rock — 121

Chapter 12: The Recipe Taken to a Professional — 128

| | |
|---|---|
| Chapter 13: Nieces and Nephews | 139 |
| Chapter 14: The Return to Rip's Basement | 150 |
| Chapter 15: Send Kerilee to Save Roger | 165 |
| Chapter 16: It's Time for Time | 178 |
| Chapter 17: Dimwitty Danger | 188 |
| Chapter 18: Who Summons Sky Woman Now? | 199 |
| Chapter 19: The First Stop in Time | 210 |
| Chapter 20: Locked in a Barn | 220 |
| Chapter 21: Return of Tip and Eberhard | 230 |
| Chapter 22: The Second Stop in Time | 238 |
| Chapter 23: The Stalker | 256 |
| Chapter 24: Third Stop in Time | 265 |
| Chapter 25: John Fisher Revealed | 277 |
| Chapter 26: No Time Like the Present | 290 |
| Chapter 27: Organizing the Village | 302 |
| Chapter 28: Basket Making | 320 |
| Chapter 29: Making the Traps | 329 |
| Chapter 30: Calling All Pukwudgies | 342 |
| Chapter 31: Night Berry Time | 355 |
| Chapter 32: The Unusual Spices | 364 |
| Chapter 33: Pukwudgie Production | 373 |
| Chapter 34: The End of Mahtantu | 382 |
| Chapter 35: Healing the Trees | 400 |
| Chapter 36: Molenhaven Celebrates | 413 |

| About The Author | 423 |
| Praise For Author | 425 |
| Books By This Author | 427 |

# CHAPTER 1: WINDMILL WITH THE SUNFLOWER DOOR

**K**erilee Oberon jolted upright. A long, slow creak echoed through the darkened library. Her heart skipped. She was alone—wasn't she?" The building had been empty a moment ago—hadn't it?

Her bare feet hit the wooden floor with a slap as she moved cautiously past the rows of bookshelves toward the front entrance, where early morning sunlight spilled in.

She had barely opened the doors and locked them behind her when she'd ducked into the break room. Now, the heavy doors stood wide open, and with them came the scent of honeysuckle… and trouble.

Someone else must have keys. She just didn't know who that was.

It was 7:10 am, and the library wasn't supposed to open until 9 am.

Strange knocking sounds came from the back of the library.

Was it haunted?

Haunted or not, Kerilee couldn't stop herself from yawning and stretching her arms. Walking back to the break room, she noticed a box of doughnuts that weren't there before. A large bite had been taken out of one that had sprinkles.

Walking back to the front desk, the scent of fresh shoe polish crossed her path.

Mysty, her feline companion with far too much attitude for someone who weighed ten pounds, perched smugly on the front desk.

"You've got company," she said in a low,

purring voice. "And you're wearing no shoes. Classy."

Kerilee ignored her, brushing her unruly braid over one shoulder as she approached the threshold. Standing just outside, as if he'd conjured himself from the misty morning air, was Sheriff Len Embrey.

"I didn't mean to startle you," he said, raising a hand in greeting." Though I see I succeeded anyway."

"Not really," Kerilee said. "I knew you were sneaking around."

"Wow, you must really have psychic powers?"

"My so-called powers are only a month old and not working that well yet."

"Then how did you know that it was me?"

"I didn't need powers to know you were sneaking around. You stepped wrong—your energy shifted the air, which is your psychic signature."

The Sheriff's smile vanished. "Plus, only you, myself, and the Head Librarian have keys and she's already here."

"Mmm. So, taking a bite out of the sprinkled donut didn't fool you at all."

"Not when you have a sprinkle or two still

hanging at the edge of your mouth."

Kerilee smoothed the front of her cardigan and forced a polite smile. "Welcome to the library, Sheriff. You also did tell me to expect you every Monday."

He stepped inside, ducking slightly beneath the carved archway, rubbing off the colored sprinkles with his thumb and forefinger.

"I forgot I said that. Anyway, I just thought I'd check in on our newest librarian. Head Librarian, Margaret Fallow, said you might be settling in faster than expected. She asked me to stop by and see how our psychic sleuth is doing."

Kerilee blinked. "She—said that?"

"She did," he said with a wry smile. "Margaret's been the Head Librarian here for decades, and she's got a sixth sense for people. Said you had potential. Said you were the kind of person Molenhaven needs—someone who could walk the line between the seen and unseen."

Mysty hopped onto the counter behind him. "You're not kidding," she said with a flick of her tail. "It's a very thin line."

Kerilee gave the cat a sharp look. "Mysty."

"Don't mind me," the cat said. "I just eat my weight in cat food everyday."

Sheriff Embrey chuckled, then lowered his voice slightly. "Margaret also mentioned you might need someone to help you adjust to the—paranormal quirks of the town. We've got more than just witches and dyslexic zombies around here."

Kerilee raised an eyebrow. "I was told this was a quiet village."

"Liar," said Mysty looking at Kerilee.

"It is," he said. "Except when it's not." He leaned against the front desk. "I've got a bit of an odd request. Ever heard of Adrienna de Groot?"

Kerilee's stomach fluttered. "The honey maker? Yeah, I've heard of her. Never met her though."

"And the Van Winkle family curse," the Sheriff added, voice dropping. "She'd like to meet you. Today. At the Honey-Making Windmill."

"Wait. Van Winkle family curse?"

"Yes."

"You slipped that in there like I wouldn't notice," said Kerilee. "How does this Adrienna know about me?"

The Sheriff cleared his throat and pointed at Mysty. Kerilee gave the cat a cold stare.

"What?" said Mysty. "Everybody in this town

knows I'm a talking cat. I walked around and said hi to everyone, what's wrong with that?"

"Sounds like you said more than just, hi."

"I told them about my new student. I told them how you're going to cleanup this town."

"How am I going to clean up this Van Winkle curse?"

Mysty didn't answer right away. "Um—We'll figure it out, you'll see."

Kerilee groaned out loud. "When am I supposed to visit this—Adrienna?"

The Sheriff made a face, saying, "Now would be good."

"How do I get there?"

"When you first drove into town, you saw four windmills. Well, hers is the one closest to the road."

"Simple enough," said Kerilee. "I guess I'll get my shoes back on."

Just before leaving for the windmill, Kerilee stopped at Margaret Fallow's office. Margaret, a silver-haired woman with eyes like sharpened quartz, looked up from a stack of town folklore volumes.

"You're off to meet Adrienna?" Margaret asked without looking at the clock.

"How do you always know these things?" Kerilee replied, only half-joking.

Margaret closed her book gently. "Because this town has rhythms, Kerilee. You just haven't learned how to hear them yet. And because Sheriff Embrey told me he was stopping by. But mostly the first thing."

Kerilee gave a nervous laugh.

Margaret stood and approached her, hands clasped in front of her. "I brought you here for a reason. This town—needs more than books and overdue notices. It needs someone who sees beyond what's written."

Kerilee hesitated. "You really believe I can help?"

"I believe you already are," Margaret said. "Now go. Just—trust your instincts."

Kerilee exited Margaret's office and was about to grab her coat when Orrin Sutton stepped through the front doors, carrying two boxes labeled with the town seal. She had seen him come and go daily since she began work at the library and her psychic senses kept picking up that he had strong feelings for a certain girl. Mysty confirmed the details and so Kerilee wanted to tease him because he was so afflicted with affection that he couldn't bring himself to talk to her yet.

"Delivery for the library," he said shyly, setting the boxes on the nearest table.

"Thanks, Orrin," Kerilee said, smiling. "Hey, quick question."

Orrin froze mid-turn. "Uh, yeah?"

Kerilee folded her arms, teasing. "So how's that girl you've been stalking?"

Orrin went bright red. "I-I don't stalk anyone!"

Mysty stretched across the counter. "He means Ruth, the Sheriff's niece."

"I do not—well, okay, I maybe… admire her… from afar," Orrin mumbled.

Kerilee grinned. "Relax, Orrin. I think it's kind of sweet."

He scratched the back of his neck. "One day, I'll gather the courage to actually talk to her."

"Better make it soon," Mysty said. "You never know what curses might sweep through town next."

Orrin offered a nervous smile before backing out through the doors, nearly bumping into a lilac-clad witch who was sliding a book into the book return bin.

Kerilee slid her car keys into her jacket pocket and glanced at the map Mysty had unhelpfully batted off the table earlier. She was new in Molenhaven—had only arrived three weeks ago from Nethertown—but the name Van Winkle had come up in whispers and glances too many times to be coincidence. Plus, Kerilee had read the book many times since she was a child.

Nethertown had its own web of conspiracies and inexplicable happenings, but it was Margaret's handwritten letter that had pulled Kerilee out—an invitation to a job and a purpose: join Molenhaven's library as a second librarian... and as a psychic sleuth.

"You sure about this?" she asked Mysty, who sat atop the dashboard, tail twitching.

"You could stay here and alphabetize cookbooks again. Or you could go meet the woman whose boyfriend might be cursed by a demon."

Kerilee exhaled. "Helpful."

She wasn't sure her psychic senses were even reliable anymore. They flickered on and off like faulty wiring. But something about the Sheriff's tone, the way he'd mentioned the windmills—plural—made her gut knot with a familiar thrill of dread and curiosity.

They hadn't made it far down the road when Kerilee hit the brakes.

A woman in a lilac-colored witch outfit walked across the street, muttering to herself. The hem of her robe shimmered in the sunlight, and her pointed hat bounced with each step. On her shoulder was a stitched patch that read: *Lily-Licking Lilac Coven*. Was it the same one Orrin nearly backed into at the library or a different one?

Kerilee turned to Mysty. "Does that... happen often here?"

"Every third Tuesday," Mysty replied without missing a beat. "She's the nice one."

"It's Monday," Kerilee said, "not Tuesday."

"So, there are purplish Witches in Molenhaven."

Kerilee drummed her fingers on the steering wheel. "The reason I bring it up is that there seems to be a lot of paranormal people about."

Mysty acted like it was no big deal. "About ninety-five percent of this town is paranormal."

Kerilee exhaled rather loudly. "Any that we have to worry about?"

Mysty braced herself for a turn Kerilee had to make. "Some play at being paranormal, but some embrace it whole heartedly. Those are the ones we worry about."

"So, we worry about those that embrace it wholeheartedly," said Kerilee adjusting the steering wheel after the turn.

"We worry about both kinds. But, let's save that conversation for another day. See, that windmill there. That's Adrienna's."

When she pulled up to the Honey-Making Windmill, it looked like something out of a fairy tale. Tall, with faded white sails and a sunflower-painted door which was open. Inside, the air was thick with the warm scent of clover honey and something electric she couldn't name.

Adrienna de Groot stood beside a table of mason jars, tan arms crossed, curls wild around her face. "You came," she said. Relief and urgency warred in her voice.

"I had questions," Kerilee replied. "And an overconfident cat."

Adrienna offered a tight smile. "We've got artifacts. Weird ones. And Roger—he won't come near them. Something happened to him when he was a kid. I think it's all connected."

"So, who's Roger?" Kerilee asked.

"Roger Van Winkle, direct descendent of Rip Van Winkle. He's my boyfriend."

Kerilee nodded. "What happened when he was a child that you think is connected?"

Adrienna hesitated to answer, but then pushed forward. "Twenty years ago, Roger's father and uncle were attacked by a giant mountain lion that was possessed by a demon. Both men were badly scarred by it. Roger's uncle nearly lost his leg and now limps badly as a result. They were trying to collect ingredients for a brew to protect them from Mahtantu, the blue demon that cursed Rip Van Winkle originally. Once Roger's mother understood that those men failed, she took off into the night to get what was needed for the brew. She vanished that night. Both men suffered recurring nightmares as a result ever since. They say the demon attacks members of the Van Winkle family every twenty years or so."

Kerilee smiled. "Thank you for being brave

enough to share this with me. Most people believe the book to be fiction."

"You do believe me, don't you?" asked Adrienna.

"I do, and I'll tell you why," Kerilee Leaned on the table. "One. Washington Irving, who wrote the book, said it was true. Two. There's a postscript at the back of the book stating that an evil spirit loved to torment man, but why was it put there? Three. The little men gave Rip Van Winkle some of their brew, but never say why. Was it the brew that made Rip sleep for twenty years, or something else?"

"Right! Too many unanswered questions," Adrienna threw in. "However, the people who have lived here all their lives seem to have the answers to all the questions except one. How to stop the curse?"

"Think of this problem as a puzzle," Mysty added out of boredom. "Find all the pieces and put it all together to get the full picture."

"Right, so show me those artifacts!"

"They're downstairs," Kerilee pointed. "Follow me."

They descended into the storage level of the windmill, a narrow staircase groaning beneath their feet. Kerilee immediately felt it—the hum of strange magic, like static against her skin. Symbols had been burned into wooden beams, and crates lined the room like soldiers. A blue mountain lion plushie, dull in the light, sat beside a black iron key etched with flames.

Adrienna grabbed the blue mountain lion plushie, "Somebody broke into the windmill and put this down here. I bet it would instantly upset Roger's dad and uncle if they should ever come here and see it. When I told Roger about it, he said it would bother him too, and he was not coming over again until it was gone."

"What about the black key?" asked Kerilee. Picking it up, Adrienna held it out in her hand. "I don't know where the key came from. It was found in Roger's pocket after a dream."

As Kerilee reached for it, her vision blurred. Fire-lit forest. Smoke. A blue demon wrapped in vines let out a scream—like trees splitting under ice, like glass cracking in reverse.

She stumbled.

Mysty darted to her side. "Hey! Easy there, medium-rare. What did you see?"

"A forest lit by blue fire and smoke," said Kerilee. "A blue man-like creature with vines all around him came through the smoke and screamed."

Adrienna caught her elbow. "That's Mahtantu, the blue demon. All these freaky things in my windmill started appearing and humming the night Roger's dreams became weird."

Kerilee took a few deep breaths to calm herself. If that was a real vision, it was her first in weeks —and it left her shaken. What if she was wrong again? What if she was just losing her mind?

Back upstairs, Kerilee nursed a cup of something herbal she hadn't heard of before while Adrienna prepared a honey tasting flight —lavender, orange blossom, clover, and one she labeled "Deseret."

"This one's rare," Adrienna explained. "The Deseret bees, which are actually from the Middle-East and no one knows how they got here, produce honey unlike anything else. I've seen these bees from Iowa to Maine. It's floral and smoky and somehow—timeless."

Kerilee tasted it, savoring the complexity. "It tastes like a warm, happy memory."

Adrienna nodded. "Exactly."

The sweetness cut through the bitterness in Kerilee's thoughts that lingered from the vision of the blue demon.

"My boyfriend's family," Adrienna said, "has been cursed constantly since Rip Van Winkle disappeared for twenty years. Roger's the latest target."

Kerilee rubbed her temples. "Disappeared? I thought he fell asleep."

"Ah," said Adrienna, "he said he fell asleep, but to everyone else in town, he disappeared."

"Two questions. Is Molenhaven the village Rip Van Winkle was from? And is there only one victim every twenty years?"

Adrienna shook her head. "No is the answer to both questions. Rip lived near here, but Molenhaven was built up later from the remnant of whatever settlement was already here. Roger's father lives in Rip's house now. As for victims, the demon makes many attempts, but he is not always successful. So, I think he makes up for it when he has the opportunity. I did some research in the library with Margaret's help, and the every twenty years thing didn't seem to start happening until the early nineteen-hundreds."

Kerilee bit her lip. "Do you really believe that we can go up against a demon and win?"

"I believe," Adrienna said slowly, "that you

were meant to come here."

Kerilee didn't reply right away. The images still swam in her mind. "And if I'm not meant for this? If I screw this up?"

Mysty snorted. "Please. You've faced worse things than ancient blue demons."

Kerilee looked at Adrienna. The woman's eyes were tired but resolute. Love and fear lived side by side in her voice. And Kerilee understood.

"Alright," she said quietly. "I'm in."

Adrienna reached for her hand across the bar. "Thank you."

Kerilee didn't have time to respond.

Adrienna exhaled shakily. "Roger's dreams... they've changed. Last night, he dreamt that he went back in time about fifty years and met a girl."

Kerilee blinked. "And?"

"She kissed him... and called him by name. But Roger says he's never seen her before in his life."

# CHAPTER 2: HE POSSESSES THE BODIES OF ANIMALS

Roger Van Winkle gasped as he woke, sweat glistening on his forehead. This repeating dream clung to him like morning mist—vivid, electric. A forest bathed in strange light. Whispers in a voice he didn't recognize. And Joanna. Always Joanna.

The girl from another time.

He sat up in the small back room of his

Uncle Sylvester's Sassafras farm, his makeshift bedroom for the past six months. The ticking of the wall clock sounded too loud in the early hush. Roger rubbed his eyes and swung his feet over the edge of the bed, haunted by the memory of Joanna's lips on his and the certainty in her voice when she said his name like she'd known it forever.

But she hadn't. Had she?

He needed answers. And there was only one person who might help—and he was just outside.

Sylvester Van Winkle loaded the last case of his current batch of Sassafras brew into the back of his pickup truck. Each case bore a logo: Rip Van Winkle asleep on a hillside, rifle resting across his elbow, an uncorked jug tipped over beside him. Sylvester tapped the image of Rip thoughtfully.

"One day, Rip," Sylvester murmured, "we'll find that missing ingredient."

He turned back toward the small office behind him. The sign by the door read *Van Winkle Sassafras Farm.* The door opened, and out stepped his nephew, Roger.

Sylvester put on his straw hat and made his way to the driver's side door. Leaning his walking stick against the side of the truck, he pulled out his keys to unlock the door—then froze at the sound of a loud crack of wood beyond the edge of his Sassafras trees.

Slowly, Sylvester looked around his truck toward the trees. He sniffed the air. Carefully, he opened the truck door as quietly as he could. He wasn't sure if he was in danger yet—but he needed to be ready. Sylvester sniffed again. Listened. Waited.

"Roger, get in the truck, please."

Roger climbed in, confused. "What's happening?" he asked, leaning forward.

Sylvester shrugged, grabbing his walking stick and laying it behind the seat. "What am I worried about? Those days are long gone," he muttered.

A distant growl froze him again. "Then again... I'm not as young as I used to be. I'd be an easy target."

Roger pointed. "Uncle—look!"

A tall pine beyond the Sassafras trees swayed unnaturally, tottering back and forth.

A deafening roar made the ground tremble. A tremendous THUD shook the trees, the earth, and the truck.

Sylvester's hands were a blur as he fired up the engine. The tires squealed until they gripped the pavement, launching them onto Route 23A.

Roger slammed the lock down and gripped the dashboard. "What kind of animal sounds like that—and makes the ground shake?"

Before Sylvester could answer, timber cracked and popped behind them. A thunderous, rapid THUD-THUD-THUD-THUD pounded through the woods. The leaves quaked violently.

Roger twisted toward his window—and saw an unbelievable sight. Dozens of pine trees shot skyward alongside them, flung like matchsticks as something colossal barreled through the forest, keeping pace.

The speedometer read sixty-five, but Roger's heart raced faster.

"Are those pines?" Sylvester shouted.

"Yes!"

"Not Sassafras?"

"No—and who cares! Can't we go faster?" Roger yelled.

Sylvester floored it. The speedometer climbed—seventy, seventy-five.

Then another roar—louder, sharper, like a

bomb detonating. A giant, nearly hairless bear, the size of a semi-truck, burst over the treetops and landed in the road behind them, still sprinting.

Sylvester and Roger screamed.

The speedometer passed eighty.

Roger twisted around just in time to see the bear closing in fast. "He's gaining on us!" he shouted, watching sweat pour down Sylvester's face.

The ground trembled beneath them. The truck rattled so violently the cases of brew in the bed bounced high, threatening to fly free.

The speedometer hit eighty-five.

With a guttural roar, the creature lunged—its massive paw slashing down through the tailgate. Bottles shattered. The truck fishtailed hard.

"Hang on!" Sylvester shouted.

Orange cones blurred past the windshield. Road workers screamed and dove aside.

Tires shrieked. The truck spun.

When they finally skidded to a stop, they were facing the wrong direction—mere inches from the bumper of a parked Sheriff's cruiser. He had just arrived after visiting Kerilee at the library.

Sheriff Len Embrey stepped out of the cruiser,

wild-eyed—and from what Roger could see, not too happy.

Roger swallowed, pulse hammering in his ears.

Len Embrey, a Native American of the Mahican tribe, surveyed the scene. Roger could hear the road crew muttering behind him as the Sheriff slipped on his sunglasses and glared through the windshield.

"Got a problem, Sylvester?" the Sheriff asked dryly.

Sylvester, still puffing and wheezing, jabbered breathlessly. "Len! Bear! Big bear! Avoided my Sassafras! Came through the pines!"

Out of breath, Sylvester said no more.

Roger sat frozen. How could any of this be real?

Sheriff Embrey strolled up to the driver's window and leaned over, sniffing.

"Well... I don't smell alcohol," he said.

Sylvester slammed his hand on the steering wheel. "Dang it, Len! Stop thinking like a Sheriff for one second—remember the old stories! Your ancestors would've believed me!"

The Sheriff's gaze sharpened. "Okay, Sylvester—if my ancestors were here, they'd be thinking:

*what the hell is wrong with this crazy white man?"*

Sylvester growled in frustration. "The stories, Len! The legends! My tailgate!"

The Sheriff peered over his sunglasses. "Tailgate?"

He straightened slowly and walked to the back of the truck, lifting a bottle from one of the busted cases before inspecting the tailgate. It sagged crookedly—slashed nearly in half by giant claw marks.

Roger saw the Sheriff's expression change.

Sheriff Embrey jogged over to where a case had bounced off the truck. For the first time, Roger noticed the shredded asphalt, gouged by massive claws.

The road crew noticed too. They pointed. Whispered.

Sheriff Embrey bent to examine the ruined road, then looked up at the gaping hole smashed through the pine wall lining Route 23A.

He began to tremble. His lips moved soundlessly at first, then louder, until Roger could make out the word:

*"Yakwawiak... Yakwawiak..."*

Sheriff Embrey stood and shouted to the crew. "NO MORE ROAD WORK TODAY!"

The workers scattered.

Jogging back, Sheriff Embrey kept chanting. *"Yakwawiak... Yakwawiak..."*

A word the local Natives used for the nearly extinct hairless bear of the region.

When he reached the driver's side window again, his face no longer angry, but filled with shock and awe, he placed a hand on Sylvester's shoulder.

"This isn't about my ancestors, Sylvester," the Sheriff said quietly. "It's about yours. Follow me."

# CHAPTER 3: THE DINER CHRONICLES

The sizzle of bacon barely cut through the mounting tension as Karyn Molen held three mugs in one hand and balanced three plates on a platter in the other. Hash browns crisped on the griddle, a bell dinged for pickup, and the Windmill Diner buzzed with the lull of late breakfast chatter. That was, until the front door slammed open hard enough to rattle the napkin holders. The regulars fell quiet.

In stomped Lester and Cobb Dimwitty, all swagger and stench, their stained hunting jackets reeking of swamp musk and superiority. The waitresses instinctively moved away from them quickly.

"Morning, sunshine," Lester called when he saw the diner's owner, Karyn Molen, tossing a muddy boot up on a stool.

Karyn didn't miss a beat. "Take your foot off my furniture or I'll serve your eggs in that boot."

Cobb chuckled darkly. "Feisty this morning. Must be the Mahtantu waking up the blood."

Karyn stiffened behind the counter, the name striking her like an old bruise. "You don't speak that name here," she muttered under her breath as she set down the mugs and plates.

"Why not?" Lester said, twirling a spoon in the sugar jar. "Folks say strange things happen when you ignore the old tales and legends. Might be time for the Van Winkles to pack it in before Mahtantu finishes what it started."

Karyn approached halfway down the counter grabbing a frying pan along the way. "Say Mahtantu's name again and I'll show you how I'm going to smash you with this skillet."

"Mantantu is—

Karyn slammed a skillet on the counter. "Enough!"

The tension simmered. Lester raised his hands in mock surrender and sauntered to a booth, Cobb trailing behind. Karyn met Lester with a warning glance. She forced a breath, steadying herself. Something was stirring these two up—and they weren't smart enough to stir on their own.

As she wiped her hands on her apron, the adrenaline faded, replaced by a prickling chill along the back of her neck. This wasn't just another flare-up. It felt too rehearsed. Like something—or someone—was scripting the Dimwittys' lines.

She thought of Sylvester's face in the hospital bed twenty years ago, pale and twisted with leg pain, and her chest tightened. If Mahtantu was rising again, the diner wouldn't be safe much longer. None of them would.

Injuries weren't even the worst part. It is how the demon destroys the lives of those he attacks. "It destroyed our lives twenty years ago," Karyn muttered to herself. "If we don't figure out how to stop it this time, it'll be another twenty years of pain."

The midday rush barely covered her need for distraction. Plates clinked, orders barked, beverages poured—but Karyn's mind ticked. In the corner booth, a hunched figure sat alone, untouched pie cooling in front of him.

That was George Tallfeather.

A Native man with deep-set eyes and weathered skin like river stone, George had once been revered—and feared—in town. He was said to have done things no white man could: called fire without matches, tracked things that left no tracks, and once calmed a rabid dog with a single whisper.

These days, he came in quietly, always at the same time, always to the same booth. He spoke little, but always listened.

As Karyn passed his table, George's hand darted out, fingertips brushing her wrist like a feather. "The wind's shifting," he rasped, voice like brittle leaves scraping pavement.

She stopped cold. "You feel it too?"

He tilted his head, gaze fixed on something far beyond the diner walls. "Been feeling it for days."

She leaned closer. "What's coming?"

His pupils flicked to her, dark and bottomless. "Payback. When the Great Spirit is ignored, those that angered him will be made to pay in pain and anguish."

She swallowed hard. "Do you mean Mahtantu?"

His lips thinned. "Sky Woman will visit."

Karyn swallowed hard. *Sky Woman. The old Iroquois legend.*

Her mother used to whisper stories of a woman who floated above the world, carrying secrets in her hands.

"She's more than a story," Karyn murmured under her breath.

A bell dinged from the kitchen. Karyn flinched, glancing back. When she turned around—George's seat was empty. A faint spiral of dust drifted where he'd been, his untouched pie still steaming.

She exhaled shakily and rubbed her arms. A shadow stirred in a nearby booth. Her father watched her quietly over his plate, knife and fork set down, his face unreadable. He'd been there the whole time, silent witness to everything.

Karyn crossed the floor and slid into the booth across from him, dropping her weight into the seat. "George Tallfeather vanished again," she muttered. "It always creeps me out when he does that."

Her father didn't answer right away. He picked up his napkin and folded it with careful precision. "You still thinking about Sylvester?"

Karyn hesitated. "I was... hoping. Maybe after all this time, things were different." She lowered her voice. "He's waited twenty years, Daddy."

Her father's jaw tightened. "The demon's still out there."

"I thought it was over."

He shook his head slowly. "It's never over, Karyn. Not when you're tied to someone Mahtantu's got his claws into."

Her phone buzzed. A heavy, ominous vibration. She swallowed the lump in her throat before pulling it out. A video message.

The thumbnail froze on the sheriff's face, his patrol car in motion. Through the back window, Sylvester's battered truck trailed close behind.

Karyn's thumb hovered. Then she tapped play.

"Hey, Karyn—Lenny here." The sheriff's voice was tight, hurried. "Listen, I'm leading Sylvester and Roger to your diner. They're shook up pretty bad. I'm hoping you can calm them down."

Karyn leaned forward, heart pounding.

"Seems Mahtantu took the form of a bear this afternoon—chased 'em down a couple miles, clawed straight through Sylvester's tailgate. They say they were doing eighty-five when it caught up."

Lenny's face turned grim. "We'll be pulling up in front of the diner in a minute or two. I just hope that demon's not following us there." He exhaled hard. "See ya."

The video ended.

A cold dread settled over Karyn's skin.

Across the diner, chairs scraped and feet shuffled. Patrons threw bills on the tables, grabbing coats, mumbling excuses. They filed out fast, nerves frayed. Even the man in the next booth left his plate half-finished and disappeared into the exiting crowd.

Her father remained. He sat back, gaze steady, hands folded over the tabletop. "You know what this means, Karyn."

She stared at the darkened windows, listening for sirens—or the roar of something worse.

Her throat felt tight as she whispered, "Yeah, Dad. I know."

# CHAPTER 4: RIP AND THE BLUE DEMON

The bear had vanished, and Sylvester had been driving behind the Sheriff for over thirty minutes, but Sylvester's hands still trembled as he stopped in front of the Windmill Diner.

The Sheriff gave a wave and his cruiser kicked up gravel as it pulled away from the Windmill Diner, taillights blinking once before disappearing over the hill. Roger sat beside Sylvester in the truck, silent as ever, but his knuckles were still white on the dashboard.

Uncle and nephew exited the pickup truck and stood for a second in front of the diner. Sylvester pushed the diner door open. The sign said it was closed, which was also a sign that something was wrong. The bell chimed overhead.

Karyn was already behind the counter, wiping down a tray with fast, nervous strokes. When she looked up and saw him, her eyes widened—but it wasn't relief that flashed across her face. It was fear.

"Please tell me you didn't lead that thing here," she said, moving quickly to the window and peering out at the treeline.

"I didn't," Sylvester said. "I don't think so."

"That's not good enough."

"I'm fine, Karyn."

"Well, I'm not fine! That demon nearly took your leg off the last time."

He lowered his voice. "It was a bear this time."

Karyn turned back, eyes narrowing. "You know the demon was a cougar last time, so it could have been a bear this time."

"I thought it was. Now I don't know." Sylvester was obviously still shaken.

She took a shaky breath, then moved closer to

him. "You remember the promise. The one we made to my parents."

"I do."

"Then we stick to it. We don't give them any reason to think we're breaking it. We eliminate Mahtantu, and we earn enough to stand on our own. Only then..."

"I know, Karyn," said Sylvester.

Karyn walked to the front window. "Lenny said over the phone that the bear clawed your tailgate. I want to see how bad it is." When she saw it, she opened her mouth and her eyes flew open wide. Tears flowed from her eyes and she took in a huge breath and screamed, *"Oh, my stars and periwinkle garters! That bear must have had huge claws! It **was** the demon."* She then sobbed in hopelessness.

The bell over the door jingled again.

Kerilee stepped inside with Mysty perched regally on her shoulder like some psychic parrot. Sylvester's breath hitched. There was something about the woman—the stillness behind her eyes, the confidence in her stride—that made him think: *She'll know what to do.*

Karyn followed his gaze to Kerilee, then nodded slightly. "Maybe it's time we ask for some help."

Ten minutes later, they sat in a corner booth, mugs of something warm between them. Mysty curled beside Kerilee, watching Sylvester with one eye half open.

"We need answers," Karyn said. "He's been having dreams. The same one. Over and over."

Kerilee leaned forward. "Recurring dreams are often the mind's way of making sense of the world, remembering something it never lived or needs to live."

Mysty yawned. "Or Mahtantu is trying to torment them every night."

"It's Rip," Sylvester said. "Rip Van Winkle. In the dream, I'm watching him—like I'm there, but not really there."

Roger, who had been silent since his arrival, looked to Kerilee with hope. "Adrienna said she spoke to you about me. I'm having weird dreams now as well."

Kerilee nodded slowly. "Would all of you you let me try something? It would help me understand more so that I can find a way to help you."

They all agreed. That night, with Mysty's guidance and Kerilee's psychic anchoring, the five of them lay down in Karyn's apartment above the diner. Kerilee whispered a meditation beneath her breath. Mysty placed her paw on Sylvester's chest. Sylvester held Karyn's hand. Karyn held Roger's hand. Roger held Kerilee's hand and Kerilee held Mysty's tail gently.

And the dream took hold.

They stood together on a mountainside high in the Catskills, the wind stirring mist through the pine trees.

Rip Van Winkle trudged along a narrow trail, his musket slung over one arm. His tattered hat flapped in the breeze. At his heels padded a scrappy mutt, tail wagging.

"Wolf," Mysty whispered. "The dog's name is Wolf."

Wolf paused near a moss-covered boulder sniffing.

"What have ye found, Wolf?" Rip asked.

The dog barked furiously at the stone.

"By the sound of the bark, it must be something large."

Rip stepped forward and sniffed the air.

"Varmint, whatever ye be, ye bear the foulest stench of any beast I've yet smelled in all my days. A skunk's odor would be sweet by comparison. Come out from behind that boulder that I may extinguish ye."

A low hum began to vibrate the earth. From behind the boulder rose a creature no man or woman should ever see.

It was shaped like a man, but twice the size and burning constantly in a blue flame. Its body shimmered as if made of smoke and fire. It wore no clothing, save for wiry vines that writhed around its loins and stretched out into the air, snapping at butterflies and birds, leaving charred flowers in their wake.

"Why call ye me foul?" it bellowed.

"Because ye stink, that's why!" Rip snapped.

"What be your name and why do ye not fear me?"

"I am Rip Van Winkle, and if ye knew my wife, ye would know a woman whose tongue is the sharpest of any woman that's ever lived—and only her do I fear."

"Then Rip Van Winkle, I am Mahtantu, and I curse ye. When next ye sleep, thou shalt never wake again. And I shall return to curse thy descendants."

"Heh," said Rip, raising his musket. He fired. The ball passed clean through the demon. Wolf whimpered and hid behind Rip.

Mahtantu laughed, unbothered, and walked away, vines snapping behind him.

Rip trembled. "Do ye feel it, Wolf? I do. The curse even now takes effect."

A voice rang out from the path above.

"Ho! Rip Van Winkle!"

A little man with a thick beard rolled a small barrel toward him.

"I hear ye, little man," Rip said, lowering his musket.

"Sleep not, Rip Van Winkle, only follow me up the mountain and partake of my brew."

"Would be a rare day that I turn away from freely offered brew. What be ye called?"

"I am called Eberhard. I saw the demon curse ye and offer ye aid."

"Lead on, then, Eberhard. I shall follow. Come, Wolf, ye coward of a dog."

The dream-watchers followed them up the slope. At the summit, a gathering of small men welcomed Rip and offered him a stool and a tankard.

"Nine Pins," Mysty said, watching the game unfold. "Old game played with old powerful magic."

Kerilee glared at the cat.

Mysty lifted a paw and turned it over. "What? I'm trying to give them subtext to enlighten them."

Eberhard raised his cup. "How dost thou find the brew?"

Rip drank. "Strong. Pleasant."

"The spices protect against Mahtantu. We brewed it as strong as we dared. We know not how much it softens the curse. Take this—" he handed Rip a folded paper. "It tells how to make it. Pass it down to thy descendants."

"I thank ye," Rip said, sliding it into his coat.

The game of Nine Pins thundered with each strike. The ground shook. Thunder cracked.

Rip stood and walked down the hill. "I am tired," he said. "Come, Wolf."

He lay down among the ferns and closed his eyes.

The dream began to fade.

But before the world could vanish, the flames returned.

Mahtantu stepped through the veil of dream, his blue flames curling hotter as he turned—not toward Rip, not toward Wolf, but toward *them.* "You do not belong here," he growled, smoke hissing from his nostrils. "This memory is *mine.*"

His gaze raked over Sylvester first. "Your nights shall stretch long." Over Karyn. "Your heart shall stay chained." Over Roger. "Your love shall slip further away with every dream."

Then his burning eyes locked on Kerilee. "And *you*—the meddler—your punishment will be worse than theirs."

Mysty arched her back beside Kerilee, fur bristling, tail lashing. "You smell worse in dreams than you do in the waking world," she snapped.

A rumble shook the mountain. Mahtantu's vines lashed toward them, snapping like whips. "I have foreseen your end little cat. Deep in the earth in a tunnel!"

The ground split beneath their feet. The dream shuddered, peeling at the edges. Kerilee planted her feet, lifting her hands as if to push the vines back with sheer will. "No," she whispered fiercely. "You don't get to curse us."

But the vines coiled tighter. Mahtantu's laughter roared, louder than thunder. "*WAKE UP!*" Mysty yowled.

Light exploded behind Kerilee's eyes. The mountain, the flames, the vines— Gone.

Karyn bolted upright in the apartment's darkened room, her breath ragged. Beside her, Sylvester gasped awake, sweat streaming down his face. Roger stirred groggily, rubbing his temples.

Kerilee sat perfectly still, staring into nothing.

"Keri?" Karyn whispered.

Kerilee blinked slowly. "He saw us," she murmured. "He marked us."

And in her lap, Mysty's tail twitched. "And next time," the cat said grimly, "He may really come after us."

# CHAPTER 5:
# THE EMPTY JUG

T he smell of sassafras and French fries lingered in the air.

Tip Van Winkle slipped through the back door of the Windmill Diner just as the sun crept over the tree line. The diner was closed for the day and he had seen Sylvester's truck out front, so he figured Sylvester would be visiting Karyn. He blinked at the sight before him: Sylvester sprawled on a bench, Karyn nursing a mug of cold herbal tea, Roger looking pale, and Kerilee staring off into space as if she hadn't fully returned from wherever she'd just been.

Mysty blinked at Tip from the counter, her tail wrapped primly around her paws. "You missed our adventure," she said. "They've just come back from a joyride through one of the nastiest dreams I've ever shared."

Tip's jaw twitched. "Sylvester's dream?"

Kerilee nodded. "Rip Van Winkle. The real one."

Tip set his hat on the counter and sank into a chair. "Then you saw Mahtantu."

At the sound of that name, the air grew colder.

Kerilee studied him. "You've seen him too?"

Tip's fingers curled against the tabletop. "He took my wife. Also slashed me across my chest." He pulled back his shirt just enough to show Kerilee the beginning of the scars that the huge mountain lion had left him with twenty years ago.

Karyn sat up. "Tip—"

"I need you to see the dream," he said, eyes wet but steady. "I need to go back there in my sleep. And I need you all to come with me. Mysty said you'd be able to help us."

Kerilee exchanged a glance with Mysty. Mysty gave a reluctant nod. "We'll do it. But you need to be in your home. Where it started."

It was still morning in the Catskills by the time Kerilee and Mysty arrived at Tip's old farmhouse. The very house that had belonged to Rip Van Winkle. Sylvester, Karyn, and Roger had driven ahead of Kerilee in Sylvester's now damaged truck.

In the guest room, Tip lay on his back, eyes already closed. Kerilee sat beside him, her fingers resting lightly on his temple. Mysty curled along his ribs, her purring deep and rhythmic. Karyn, Roger, and Sylvester took up the rest of the floor space. Roger was a small child when the real event happened and he had only seen a small part of it. Though a young man now, he eyes couldn't hold back the tears and his chest heaved painfully over and over.

The dream took hold.

A soft knock sounded at Tip's front door. He was already dressed in his denim overalls and shirt and looked much younger, a large wooden basket slung over one shoulder.

He peered through the peephole. A younger Sylvester. And a younger Karyn.

He opened the door.

"Are they really out there, Sylvester," asked Tip.

"They're up the hill," Sylvester said, eyes bright. "Just like Eberhard said. Bright as day."

"Then we're finally going to fill that jug," Tip smiled.

"Yes we are, Tip, tonight's the night!" Sylvester smiled back.

"Are you ever gonna let me by, Sylvester?" came a woman's voice behind the man. Sylvester stepped aside revealing a lovely curly-brown-haired woman who looked at him adoringly.

"Sorry about that, Karyn." She gave Sylvester a loving tap on his chest and then a quick elbow to Tip's gut as she stepped into the house.

"She's a firecracker," Tip giggled as he grabbed his belly.

"What's going on?" whispered another woman's voice.

Karyn, Sylvester and Tip looked into the house to see Tip's wife tying the belt to her white satin robe as she came toward them.

"LeeAnn," Karyn grinned and threw her arms around her. LeeAnn hugged her back. "Come outside and look at them with us."

"Look at what?" LeeAnn said as she followed Karyn out the door.

Tip and Sylvester stepped out with them. Karyn pointed up the hill from the Van Winkle's thatched roof house towards bright lights on the hillside.

"Aren't they beautiful?" Karyn asked. "They look just like diamonds shining in the night."

LeeAnn's mouth opened wide as did her eyes. She reached up with both hands to grab her long blond hair. "You mean…it's true? Really?" she said.

"I've always told you it was true, my love," Tip said giving his wife a hug.

"I know, but now I get to see it." LeeAnn said kissing him back.

"We'd better pick them before they go out," Sylvester said pointing up the hill.

"Where you going, daddy?" asked a small little voice. The four adults turned and looked down to see five year-old Roger Van Winkle looking up at them.

"Uncle Sylvester and I are going to pick night berries," his father answered.

"You are supposed to be asleep," LeeAnn said picking up her son.

"I want to go too," little Roger said. Tip chuckled and nudged his brother. "See, he already wants to take on the responsibilities of a Van Winkle."

"You've got a fine boy, Tip," Sylvester said with smile.

Tip kissed his young son on the head and took him from LeeAnn. "Roger, its dark outside and I don't want you to get lost. Instead, let's go inside so I can show you something." Roger rubbed his eyes and nodded.

"He's so cute," Karyn said as everyone moved back into the house.

Tip brought Roger up to the fireplace mantel where sat Dutch knickknacks like Delft blue plates with pictures of windmills on them; wooden shoes; and an old jug.

Tip reached out and touched the jug gently, "Roger, this old jug is the very jug that Rip Van Winkle drank out of while he watched the little men bowl nine-pins."

Roger smiled, "Momma read me the story."

"Yes," Tip continued, "In the book it was called a Flagon, but that's an old word we don't use anymore. Nowadays we call it a jug." It had a small handle near the neck and had the look of stone with a large cork stuffed into its wide mouth.

Roger pointed and said, "Jug!"

"That's right," LeeAnn said reaching out for her young son as Tip handed him over, "but the jug is empty and your daddy and Uncle Sylvester have to go pick the berries so they can make the drink that fills it up."

"So you go back to bed and I'll see you in the morning," Tip said as he and Sylvester made their way to the door waving goodbye. Roger waved back until his father and uncle stepped out and closed the door behind them.

"Shouldn't take them long," Karyn said. LeeAnn sat in a rocking recliner and began rocking her son to sleep. "I hope not. Wasn't Len going to help them?"

Karyn made herself comfortable on the couch. "He got called in to work," she said as she began drifting off to sleep.

"You know how important his job is to him."

"Yeah," LeeAnn said as she saw little Roger's eyes close, "first Native American to wear a badge in Greene County, New York."

In the distance, a pair of men screaming froze LeeAnn's blood.

Karyn sat straight up. A frightening blue light flashed like lightning across the windows followed by a deep throated growl.

"Karyn, put Roger in his bed," LeeAnn said shaking. Karyn took Roger and carried him to his room and placed him under his covers. She could hear a door opening in the kitchen and LeeAnn rummaging around. Quickly leaving and closing the door to Roger's room, Karyn turned to see LeeAnn loading a pump-action shotgun.

"You can't kill it, LeeAnn!" Karyn pleaded.

"Those screams were our guys, Karyn," LeeAnn said shaking. "I've got to do something."

"I know, but the thing doesn't die. It can't be killed. Maybe we should—"

A loud piercing growl from a mountain lion interrupted her followed by Tip and Sylvester wailing in pain.

Odd sounding footfalls approached the house. Karyn ran to the front door and pulled it open as LeeAnn leveled the shotgun toward the door.

Sylvester, bleeding from one leg and crying hopped to the door carrying Tip in his arms. Tip was bloody also and was jabbering away.

"Burning man… fire, fire… blue with vines… nose burns… throat too… awful smell… sharp, sharp claws… he grew, he grew! Puma…"

Tip continued to jabber.

LeeAnn set the shotgun against the wall as Sylvester set Tip in the recliner.

"Puma got my leg," Sylvester said painfully to Karyn. "I'll need an ambulance."

"I'll call," Karyn said frantically and darted to the kitchen phone.

LeeAnn started to cry and she knelt to hold her husband. "I thought it was just an Indian Legend," she said.

"No," Sylvester managed taking an iron from the side of the fireplace to use as a crutch. "It's just as real as the rest. It doesn't want us picking the berries."

"Did you get enough?" LeeAnn asked through her tears.

"Tip threw his basket at it when I was attacked," Sylvester said looking into his basket. "I don't have enough in mine."

Karyn hung up the phone. "The ambulance and a deputy are on their way!"

"I failed you LeeAnn," Tip sobbed, "I failed little Roger, too."

"Not enough to protect the family?" LeeAnn asked.

"No," Sylvester answered. "I must have lost most of mine when I was knocked down."

A car could be heard pulling in the driveway. Karyn rushed again to the door and peeked outside. "It's Len," she said as a distant siren wailed.

"And that should be the ambulance."

Deputy Len Embrey, a tall young Native American, took off his hat as he nodded to Karyn at the door.

"What's happened?"

"It's bad, Len." Karyn started trying to hold back her emotions. "It got them both."

The young deputy saw the blood trail on the floor and Tip mumbling in the recliner. Sylvester hobbled over as Len looked down at his leg.

"What the hell, Sylvester! What did I tell you about protecting yourself."

"I thought the baskets had enough Sassafras woven in it," Sylvester said wincing in pain.

"Obviously not quite enough," the deputy said looking at the wound. "You need a tourniquet on that leg."

Karyn looked around the room. "LeeAnn?"

"LeeAnn, we're going to need something for a tourniquet!" the deputy said looking around.

"She's gone," Sylvester said worriedly.

"The shotgun's gone!" Karyn exclaimed.

"The basket too," Sylvester pointed out just as the EMTs came in the door.

"Night berries," Tip mumbled. "Foul blue demon...burning man..."

Len turned to the EMTs. "Take care of these men; I've got to chase someone down." Then turning to Karyn, "What was she wearing?"

"A white satin robe," she replied. Deputy Len Embrey charged through the kitchen and out the back door. Scanning the hillside above the Van Winkle home, Kerilee and the other dream watchers saw LeeAnn's form distantly running over the horizon with the moonlight glimmering off the back of her robe. Len tore after her pumping his legs and arms harder than he had ever known before. He knew Tip and Sylvester were lucky to survive their encounter with this evil Manitou, but LeeAnn was not likely to get the same odds. Her only hope was if it left the area long enough for her to get the berries.

However, just as Len reached the top of the rise where he last saw her, they heard LeeAnn scream and saw the bright blue blaze of fire in the distance. Plunging down the path toward the evil light, they heard a shotgun blast followed by another scream from LeeAnn. The blue light went out.

Len pulled his flashlight from his belt and switch it on. He came across the basket first and picked it up. Ten yards further was the shotgun, but no LeeAnn. As he lowered his flashlight, Everyone registered that he was standing inside a pawprint more than five feet wide. Len stood there in the dark, heart thudding.

"She's gone," he whispered.

Those watching the dream stood at the site of the attack. The ground was scarred. A paw print the size of a small car dented the earth.

They saw the basket. Crushed. Empty.

They saw the shotgun, warped.

Kerilee touched the pawprint. "This... wasn't just an animal."

"No," Tip said. "The Manitou possesses them. But it can't stay inside one body too long."

"Manitou?" asked Kerilee.

"A native word for evil spirit," said Mysty.

Tip's breath hitched. "The animal dies, and Mahtantu sheds its skin."

Kerilee looked up at the mountains. "Where did he take her?"

Sylvester, could barely speak through his emotions, "He took her to Devil's Grasp."

Roger's eyes leaked like a faucet with a bad washer as he stared at the pawprint. Kerilee put her arm around him and pulled him close. "Let's go back," she whispered. "He's carried enough of this for one lifetime."

# CHAPTER 6: SAME DINER, DIFFERENT DAY

Eberhard sipped slowly, the cup nearly dwarfing his small hands.

"You've been quiet," Karyn said, lowering her voice.

"I speak when words are useful."

Karyn scowled. "Useful would be telling me how to stop this thing. The curse. Sylvester almost died twenty years ago."

Eberhard peered at her through cloudy eyes. "You've seen the dream? How the demon will do anything to prevent anyone from ending this curse?"

"Yes."

He smiled. "Then why did you come to me?"

Karyn leaned in. "Can you help or not?"

"It will take a mountain of materials and an army of people to do it, but first we need Kerilee. And the cat."

"Of course. The latest psychic and her talking sidekick." Her jaw tightened. "I like the girl, you understand. I just don't like depending on others."

"No one does. That's why they fail."

He sipped again, then murmured, "I noticed you did not say anything about liking her cat."

Karyn spun her cup slowly on the table. "I feel like it judges me."

Eberhard shrugged, "Mysty's purpose is to evaluate situations and to guide. That does require making proper judgments."

The bell jingled.

Kerilee Oberon stepped into the diner, Adrienna at her shoulder. But Kerilee paused in the doorway, wondering where Mysty had gone off to.

Karyn waved her over. "Grab a seat. You just missed the swamp rats. Those Dimwittys were here again."

Adrienna smiled as she walked in and sat. "Karyn, tell Kerilee how the Dimwitty's played a part in my meeting Roger."

"Oh that," said Karyn as Kerilee sat next to Adrienna.

"Roger became my delivery boy right when he got his driver's license. He had just come back with a bunch of supplies when Lester and Cobb Dimwitty and their buddies came over to harass Roger. They were shoving him around when this little scrap of a girl I'd never seen before shows up and starts sticking up for Roger. When the Dimwittys turned on Adrienna here, she gave Lester an uppercut that lifted him into the air and onto the sidewalk. That made the whole bunch of them run off."

"Without saying a word to Roger after that, I walked into the diner and sat at the counter," said Adrienna. "Roger brought in the boxes from the truck and just stared at me."

Karyn laughed. "I had to tell Roger to say thank you to her. He was in shock that anyone would take on a Dimwitty for him."

"I did it because I thought Roger was real cute and because I really hate bullies."

Kerilee pointed to Adrienna's hand. "Didn't that hurt?"

Adrienna looked at her hand. "It sure did and Karyn, here, had to put a lot of ice on it and wrapped it up, but to win Roger, it was sure worth it."

"Oh—Margaret Fallow came by earlier," Karyn added, lowering her voice. "Said to tell you to check the library archives tonight."

Kerilee frowned. "Did she say what for?"

Karyn shook her head. "Just said you'd understand when you saw it."

Kerilee frowned. "I'm suddenly sensing someone—Eberhard?"

"In his booth, waiting like a crypt keeper." Karyn nodded toward the far booth. "George Tallfeather came by again earlier. Real quiet. Didn't stay long—just sat a while, then left."

Kerilee glanced toward the corner. "Are they both always so—interesting?"

Karyn gently cleared her throat. "In this town, everything is weird or interesting. You are the most recent person to move here. Just wait, you'll experience things in this town you never thought could happen."

Kerilee drummed her polished finger nails against her mug. "The Sheriff said something to me about dyslexic zombies and other strange things."

"Speaking of strange, I want to show you two my memory wall," Karyn told them.

Karyn led Kerilee and Adrienna to a booth by the far wall, where a dozen Polaroids clung to an old corkboard. Most were faded snapshots of picnics, Fourth of July floats, snow-dusted porches.

One showed a younger Karyn and Sylvester in the woods—she was laughing, hair caught in a breeze; he was staring at her like she hung the moon.

Karyn stared at the photo, the moment blurring into memories—they became a shared vision in Kerilee and Adrienna's minds.

*Karyn and Sylvester had been chasing frogs near Rushing Hollow. Sylvester had scooped*

*one into a mason jar and handed it to Karyn with a grin that crinkled the corners of his eyes.*

*"If I catch a shooting star, will you marry me?" he had teased.*

*"You better," she had said, brushing her wind-blown bangs out of her face. "Because I already made the wish."*

*Karyn's throat tightened as the memory switched to the day she learned about the Van Winkle curse. The memory rose like an old song.*

*Sylvester's voice echoed in her mind— softer, younger, laced with unease.*

*"My dad told me last night... it started with Rip. Not the story in the books. The real Rip. He made a deal in the mountains. A deal he didn't even know he made."*

*She remembered frowning at him. "A deal?"*

*"The sleep wasn't the worst part of the curse," he'd whispered. "The worst part was what came after. Rip woke up... but things started happening to his relations. The demon felt cheated that Rip only slept twenty years. So, he began cursing every Van Winkle in whatever way he felt like. He snatches people and turns them into trees above the falls."*

*Even now, twenty years later, she could still hear the haunted edge in his voice.*

*"Mahtantu's been following us ever since, and Rip had looked for ways to stop him," Sylvester had said quietly. "And someday... one of us will have to finish what Rip started. Or we'll never stop running."*

Karyn opened her eyes, staring at the photo,

his words curling inside her chest like an old ache.

Kerilee took a breath as she recovered from the vision, "You've been through a lot of heartache."

Karyn swallowed and offered a small, bitter smile. "You don't know the half of it."

Adrienna didn't press but kept her eyes on the image. "I don't think love like that disappears. Maybe it waits. Maybe it finds a way."

Kerilee gave Adrienna a look that blended gratitude and surprise. Then she turned back to Karyn. "Do you regret it?"

Karyn didn't hesitate. "Only every single night."

Silence settled.

Mysty, unnoticed until now, curled beneath the table. Karyn looked down absently. "You think I messed up, don't you?" she whispered.

"No. I think you did what most humans do —choose love, mess it up, and keep hoping the person you chose wasn't a mistake."

Karyn smiled faintly. "I never stopped loving him."

"I know."

"Do you think he ever stopped loving me?"

Mysty lifted her chin and gave a slow blink.

"He never even slowed down."

Karyn's throat tightened. She gave the cat one last look, then stood.

As Eberhard appeared beside them, Kerilee's breath caught. A pulse thrummed in her chest, distant thunder rolling beneath her skin. The air bent, shimmering faintly around his small frame.

And then she saw.

Not with her eyes, but behind them. A flash of wings—neither feather nor bone, but light spun into iridescence. An ancient presence cloaked in mortal shape. His gaze carried the weight of eternity.

The knowing struck like a silent bell. Not old—timeless. One of the Cherubim. A sentinel meant to bar the way to the Tree of Life until man was free of sin. She wasn't just dealing with ghosts and visions now. This was biblical. Cosmic. And he was standing right next to her.

He looked at her, and in his lined face, she saw echoes of fire and grief and duty too heavy to name.

Eberhard gave a soft, almost pitying smile.

"Took you long enough to register who I am."

Karyn raised an eyebrow. "Have you met?"

Kerilee shook her head. "Is this Eberhard?"

"I meet everyone. Eventually." He turned to Karyn.

"You've seen the threads," Eberhard murmured. "She'll find the needle."

Kerilee's frown deepened. "Then I'd better learn to sew."

Eberhard chuckled softly. "Fate's already stitching."

Kerilee huffed in confusion, "What are you trying to say?"

"The Sky Woman," Eberhard grinned. "She isn't just a legend. She's the only one left who can call the jailers back. The ones who exiled Mahtantu to this world in the first place."

Karyn tapped Kerilee's arm, "Maybe the library might have more information on the Sky Woman."

"The truth doesn't sit on a shelf," Eberhard said. "It sits behind your eyes."

Kerilee said nothing, but something shifted behind her gaze.

She'd find the answers. Even if it meant facing Mahtantu. Alone. But one question lingered:

Eberhard.

"If you have a question, Kerilee, then ask it," Eberhard's voice echoed—not aloud, but *inside her mind.* She jolted slightly.

Kerilee gasped quietly, her eyes wide. He hadn't spoken aloud. But she heard him clearly—as if he whispered from behind her thoughts.

"Eberhard," she asked silently, unsure how she knew how, "why would a cherub be here..."

"Surely you've heard—we're facing a blue demon. Not one of the ones cast down with the devil. No... his own people got so fed up with him, they came here, dumped him, and left him behind."

Eberhard's head turned. His gaze locked on the frost-etched window.

"He's listening and watching," Eberhard murmured.

A chill swept the diner. The lights flickered. Outside, the wind howled.

"You don't mean... watching *us?*"

Eberhard turned, meeting Kerilee's gaze. "The Blue Demon *is* watching."

Mysty's tail twitched sharply. "Then let him watch."

Kerilee exhaled slowly, her fists clenched

beneath the table. "He's not the only one paying attention," Kerilee said. "Someone else is coming to find us."

# CHAPTER 7: THE MAHTANTU EFFECT!

The butterfly never saw it coming.

It had flitted through the cracked window of Adrienna's windmill, golden wings catching the dusty sunlight. Mysty tracked it from her perch on the windowsill, tail twitching, her whiskers quivering with unease. A shiver of wrongness rippled through the air, faint but undeniable—like a key turning in a lock that shouldn't exist.

Kerilee was murmuring to Adrienna near the

base of the windmill, running her fingers along one of the etched panels they'd found last time. Mysty flicked her ears toward the girl. Adrienna's presence always came with an odd hum—subtle, like the vibration of moth wings near candlelight. But now it was louder. Stronger. Mysty narrowed her eyes. The girl didn't know it, but she was stirring something dormant inside her. Psychic threads woven into her soul were beginning to flutter. 'She doesn't know what she is yet,' Mysty thought. 'But it's waking up. And that's either very good—or very bad.' Mysty leapt down with practiced grace and padded silently across the floor, following the butterfly's drift toward the vines carved into the old stone.

The carvings shimmered.

Not with light, but with awareness.

Just as the butterfly passed over a delicate spiral of leaves, a jagged tendril *moved*—snapping out in a flash and catching it midair. There was no sound. Just the sudden absence of flight.

Mysty's ears flattened. "Mahtantu's vines," she muttered to herself, "they strike at joy."

She padded to Kerilee's side just as the young woman pressed her hand to the etchings. The moment flesh met stone, the air shifted again.

Kerilee gasped.

Mysty's mind linked to hers in a rush of shared vision—loud booms, the neighborhood shaking, something was coming, something huge.

Kerilee was never more frightened in her life.

Mysty growled, low and guttural, severing the connection with a paw to Kerilee's leg.

"Don't touch them again without protection," she hissed. "Those aren't drawings—they're bait."

Kerilee, still pale, nodded. "What did I see? What was that?"

"I'm not sure," Mysty said, ears swiveling. "Something yet to come. Hope it isn't watchers."

"What watchers?"

"You don't know about watchers?" Mysty said with surprise. "Didn't you ever watch weird videos on the Internet about supernatural events and beings?"

Kerilee blew out a breath. "Not really. Weird didn't happen until you came along."

Mysty grinned. "Good. That way you weren't tainted by videos that are filled with misinformation. I can teach you how to do weird properly."

Back at the Molenhaven Library, the dusty hush felt almost too still. Kerilee flipped through the folklore archives that Margaret had wanted her to do, while Mysty prowled the baseboards, whiskers twitching.

Then she stopped. Something called to her.

She swatted at a seam in the wall beneath the shelf labeled *Founding Myths & Miscellany*. A hollow thumped behind the panel.

But before she alerted Kerilee, Mysty's whiskers twitched again. Something else was nearby—old, metallic, forgotten.

She followed the scent to a panel behind the archive cart. The corner of the panel was missing and some fabric could be seen within. Mysty reached in with a claw and began pulling the fabric out. Beneath a dusty wool tarp lay a long object crusted with rust. Mysty clawed at the fabric until it fell away, revealing a rifle—barrel warped, stock cracked with age.

Kerilee turned and gasped. "Is that...?"

"Rip's," Mysty said. "I'm sure of it. It looks like the one we saw in Sylvester's dream. Left outside when Rip slept for twenty years. Rain and time did what the curse didn't."

Kerilee knelt beside it, reverent. "Even though I see his musket here. I still find it hard to believe that he just... slept for twenty years."

Mysty flicked her tail. "And this waited for him to wake up, rotting."

They stared at it a moment longer, then turned back to the panel.

"Here," she said.

Kerilee pried the panel loose. There was an old book which Kerilee put on white gloves for to remove carefully. On the cover, written in large calligraphy letters was : The Journal of Rip Van Winkle.

Margaret Fallow watched from a distance with a smile on her face. The librarian's gaze lingered not on the journal, but on Kerilee—on the way her fingers trembled as she turned the pages, the reverence in her posture. Margaret had seen hundreds of patrons open books, but few had ever *read* them the way Kerilee did—with her whole soul listening.

'She's the right one,' Margaret thought. 'Rip's truth never belonged on a shelf. It needed a seeker—someone with the gift, and the heart to carry it.'

She noted the moment Mysty brushed her tail against Kerilee's ankle, a silent signal only a magical creature would give. That bond between them—of soul and intuition—meant everything.

Kerilee scanned pages aloud. Rip's entries were rough but legible:

> *"He wanted me cursed forever, but they called to the sky and my curse was reduced to only twenty years."*

> *"The brew—Eberhard called it the Brew of Mercy. Sassafras, fire-root, spicy roots, and night berries."*

Mysty sat perfectly still. "The brew softened the curse. That means it can be undone."

Kerilee's eyes widened. "Or redirected."

Farther into the journal, the tone shifted. Later entries were filled with fragmented dreams and drawings of shadowed faces.

One name appeared again and again: *John Fisher.*

Kerilee read: "He came to me while I slept. Said he knows the beginning. Said he knew where the demon came from."

Mysty's tail twitched. "He watched us too. Last time. In the dreamscape."

Kerilee flinched. "Who did?"

"John Fisher."

Kerilee laid a clean piece of paper to mark her place in Rip's journal. "I felt that in the diner."

Mysty narrowed her eyes. "He's not a mortal human."

Kerilee looked up. "What do you mean?"

"There are only five men before in Earth's history who've been blessed as he has. They walk the line between flesh and glory."

Kerilee's mouth went dry. "How do you know this stuff, Mysty?"

"Oh, I guess I said too much, again," Mysty said.

"C'mon now," Kerilee said. "Just tell me what you can."

Mysty's gaze darkened. "Six men in history. Also in scripture—not just the ones you know, but hidden ones too. They were granted something very special."

Kerilee's breath caught. "Six men?"

"Maybe seven. We know the names of three," Mysty said quietly. "Prophecy tells us what they'll do one day…"

"Could they help us with the curse?"

"Only if it is part of their mission," Mysty said. "I cannot say anymore at this time."

"It sounds very interesting, but why keep it secret?"

"It is part of an ancient law. If you seek to know it, you must search it. When you have found all you can, only then will the rest be given to you. In other words, 'Seek, and ye shall find.'"

"There is so much I have yet to learn about the paranormal world," said Kerilee.

"That is true, Keriliee," Margaret's voice said in the distance. "Yet, it is the path I very much hope you'll follow."

"Margaret, do you know much about this ancient law and these six men?" asked Kerilee.

"I know many books of true scripture. I know the names of the three others that people know. But even at my age… I don't know everything. However, if you want to know about them, I'll email a list of items to read."

Mysty snickered into her paw. "You'll be reading that list for years."

"Gee thanks," said Kerilee.

Kerilee copied the passage about the Brew of Mercy into her notes while Mysty wandered to the library window. A breeze stirred the branches outside. Nothing unusual.

Until the trees shivered.

Not swayed. *Shivered.*

A squirrel bolted across the lawn and then stopped mid-stride. It trembled once... then collapsed, its life drained by a gray vine creeping across the grass.

Mysty yowled and leapt down.

"Kerilee. We have to move. *He's here. The blue buttface is outside the library.*"

Kerilee didn't argue. She shoved the journal into her satchel and nodded. "We'll go to the diner"

Mysty had faced a demon once before—not Mahtantu, but one nearly as cruel. A demon who cursed someone Mysty loved, turning their laughter to dust and their memories to smoke. Mysty had fought with all her power to undo it, but it had taken too long. The pain of that failure still burned under her fur.

That's why Kerilee mattered so much.

That's why she would do *anything* to keep her

from facing the same fate.

A breeze burst through the library window—and with it, a piercing, unnatural scream echoed across the ether. It shook the panes and made the dust on the shelves dance.

Mahtantu.

He'd heard Mysty's insult. He *felt* it.

Mysty grinned wickedly, ears back, and yowled out the window, "Poor baby. Did the blue buttface not like that?"

"Outside, the vines writhed, clawing across the lawn."

"I'm not your ordinary cat," she added, tail swishing high. "I'm the kind you *can't* curse, demon."

Another scream split the sky, more furious this time. Mysty let out a snort and laughed so hard she nearly rolled off the sill.

"They never learn," she purred, satisfied.

The journal jumped from Kerilee's satchel onto the floor. The final pages flipped of their own accord. A map pulsed faintly in the margin, drawn in faded ink. It showed mountains to the north—and carved into the edges, the same symbols etched in the windmill walls.

"Don't touch those pages!" shouted Mysty.

Kerilee kept her distance. "I see the same symbols as—

at Adrienna's place. They're bait remember."

"Why do I get the feeling that journal is fake?"

Mysty slowly approached the journal, sniffing as she went. "I think this was planted by the Dimwittys."

# CHAPTER 8: ADRIENNA'S DISCOVERY

The hive cracked open with a sound like snapping bones.

Adrienna winced as sticky golden honey oozed over her gloves, flooding her nose with warm sweetness. She scraped a chunk into the mason jar balanced beside her elbow and muttered, "Sticky little beasts. I swear, if one more stings my butt, I'm torching the whole barn."

She'd only meant to harvest a few frames today. But something felt... wrong. The air was too still. Too thick. Like the forest itself was holding its breath.

Then her knife struck something solid.

**Clink.**

Adrienna froze. "That's not beeswax."

She brushed aside wax and crumbling honeycomb, squinting into the sticky mess. Wedged inside the lower edge of the frame was a small, tarnished object. Metal. Oval. An old charm, no bigger than a walnut, bound in copper wire etched with strange looping marks.

Her breath caught.

It pulsed faintly in her palm. Not heat. Not light. Just… a hum. A low, unsettling buzz, like a bee trapped inside her bones.

A chill crawled down her back.

Part of her wanted to tuck it in her pocket—like some secret treasure. But another part—the one that remembered the look in Kerilee's eyes when they'd last talked about curses—wanted to yeet it straight into the woods and run.

She didn't.

Instead, she whispered, *"Sky Woman?"*

No answer. Only the air growing colder, the hairs on her arms standing up.

"…Why did I whisper that?" she muttered.

A few minutes later, Adrienna stood at the tin sink outside the barn, scrubbing honey off her hands. The charm was wrapped tight in a rag, hidden snug in her pocket.

She leaned over the basin, cupping water to her face.

Then she looked down.

The reflection staring back *wasn't hers.*

A young girl with long, dark braids gazed up at her from the water. Serene. Luminous. Eyes shimmering with something Adrienna couldn't name—tenderness, maybe. Or memory.

Adrienna's breath snagged in her throat.

The girl's lips parted like she might speak. But before a word formed, the image dissolved into ripples.

Gone.

Adrienna stumbled backward, heart pounding. "Okay... what the hell was that?"

Her hand reached into her pocket for the charm.

Roger's voice echoed in her head—he'd described her. That girl. In his dreams. Joanna.

A cold knot tightened in Adrienna's gut. A flicker of jealousy bloomed, sharp and sudden. Irrational? Maybe. But real.

"Seriously?" she muttered. "You're dreaming about *her*?"

She stared at the charm. Suspicion flared. Was this charm planted? A warning? A trap?

She clenched her jaw, squared her shoulders, and turned toward town.

"Fine. I need Kerilee. Like, now."

The woods felt wrong.

Not just dark—*watchful.*

Adrienna moved quickly, hand jammed in her jacket pocket, feeling the charm vibrate faintly against her palm like a hidden heartbeat.

"Fantastic," she muttered. "Haunted jewelry. Love that for me."

A branch snapped nearby. She spun, heart in her throat.

Then—out of the shadows padded Mysty, black fur glinting silver under moonlight.

"Evening," the cat purred, voice smooth as silk. "You smell like bees and *Dimwitty stink.*"

Adrienna froze. "What the heck does that mean?"

Mysty circled her ankles, tail flicking. "The forest hums tonight. Some trees whisper truth. Others..." Her ears flattened. "They're listening for *him.*"

Adrienna swallowed. "Mahtantu?"

"The very one." Mysty's golden eyes gleamed. "And that charm you carry? It stinks of lies and lost things."

Adrienna pulled it out, still held tight in her grip. "I thought... it might be blessed. Sky Woman—"

"Sky Woman doesn't bind her gifts in copper wire," Mysty snapped, stepping back. "And she doesn't hide them in a hive where death gathers."

Adrienna frowned down at it. "Then what is it?"

"I don't know." Mysty's voice softened, almost reluctant. "But I don't like it."

Neither did Adrienna. Not anymore.

She tucked it deeper in her pocket and stared into the dark trees. "Why does this feel like the part of the horror movie where the audience yells at the girl to turn around and she doesn't?"

Mysty's tail twitched. "Because it is."

Adrienna sighed. "Cool. Love that."

Night fell like a curtain.

Back at the Windmill Diner, Adrienna sat on the porch steps, elbows on her knees, the charm laid out on a folded napkin beside her. Mysty sat nearby, eyes narrowed, tail wrapped primly around her paws.

Kerilee hadn't showed up yet.

The insects had gone quiet.

A sound floated from the woods—low and distant. Not a growl. Not a howl. Something stranger. Like a cough. Like a beast trying to clear its throat.

Adrienna tensed. "Oh no. Nope. Not a fan of whatever that is."

Then—

A shape moved in the moonlight.

A stag.

Massive. Still.

Its body loomed just beyond the tree line, antlers spreading impossibly wide, tangled with glistening black vines. The vines shimmered, slick and wet-looking, as though they'd just crawled free from a nightmare.

It turned its head. Slowly. Deliberately.

Its eyes locked on Adrienna.

They glowed faint blue.

Adrienna's breath caught. "That's… that's not normal, right?"

The stag stared a beat longer. Then—

It vanished. Like mist. Like memory.

Adrienna's heart thudded wildly in her chest.

Beside her, the charm rolled an inch closer to her hand.

Adrienna jerked her hand back. "Not today, Satan."

Mysty's ear flicked. "Good instincts."

Headlights cut across the lot. A white car pulled up. Kerilee stepped out.

Adrienna shot to her feet, snatching up the charm. "Kerilee!" She ran to meet her, shoving the wrapped object toward her. "This is the charm I told you about. It feels wrong. It's messing with my head."

Kerilee took it carefully, frowning as she turned it over in her hands. "Oh, yes. This is bad. How do you feel now that you're not holding it?"

Adrienna swayed a little. "Honestly? Tired. Strained. I'm glad it's gone—but part of me wants it back."

Kerilee studied her closely, moving the charm away, watching Adrienna's eyes follow. "Okay. Not great. I think someone's using this charm to influence you."

Adrienna pressed a hand to her stomach. "I feel nauseous."

"When did you last eat?" Kerilee asked.

Adrienna squinted. "Last night... I think?"

Mysty padded between them. "Well, good thing we're standing in front of a diner. Let's feed the human before she passes out."

Kerilee nodded, looping an arm around Adrienna's shoulders. "Come on. Let's get you something to eat."

Kerilee gently steered Adrienna through the diner door. The warm scent of grilled food wrapped around them like a hug.

Behind the counter, Roger flipped a burger on the griddle. He froze mid-flip on the adjacent burger when he saw them.

Adrienna stiffened. "I don't want to see him." Her voice wavered. "He's been cheating on me. With that Joanna girl."

Roger blinked, burger still suspended on his spatula. "Uh—"

"Don't drop it, Roger," one of the waitresses called without looking up.

Adrienna's eyes narrowed. "I'm going to strangle her with those dark braids."

Roger slowly lowered the burger onto the bun. "I... what?"

Mysty hopped onto a stool. "Roger, make her whatever her favorite meal is."

Roger swallowed. "Okay. One grilled shrimp

salad, coming up."

Kerilee guided Adrienna into a booth. "You're a little... combative tonight."

Adrienna crossed her arms, glaring at the waitresses. "How many of them has he cheated with?"

"They're high school girls," Kerilee said gently.

"They're still girls!" Adrienna snapped.

Kerilee leaned forward. "You told me yourself—Roger's cursed. He can't help seeing Joanna in his dreams."

"He could tell her to buzz off. *Something.*" Adrienna stabbed the table with her finger.

Kerilee sighed. "I think that charm's really messing with you."

A waitress set down a salad and utensils. "Here's your shrimp salad. Someone will bring your water in a minute."

Adrienna unrolled the napkin, grabbed the fork, and pointed it accusingly at the waitress. "How many of you are dating Roger?"

Kerilee reached to lower the fork. "Adrienna—"

Adrienna jerked it away. "Back off, Kerilee! I need answers!"

The waitress raised both hands. "None of us.

Ms. Molen made it a rule. No dating Roger if you want to keep your job."

Adrienna stared. Slowly lowered the fork. "Karyn's a good woman." Her voice wavered. "What's wrong with me?"

"Either the charm's cursed," Mysty chimed from beneath the table, "or your cycle's coming."

Adrienna groaned.

"Either way," Kerilee said soothingly, "you've got food now. Eat."

Adrienna dug in. Roger had piled extra shrimp and a few special toppings she loved. Between bites, her eyes fluttered closed.

A waitress slipped by and left a tall glass of water without a word.

Kerilee watched as Adrienna's shoulders slowly relaxed.

Roger brought over a dessert—a brownie drowned in fudge and chocolate chips.

When Adrienna had demolished it, Kerilee asked gently, "Okay. How do you feel now?"

Adrienna leaned back, sighing deeply. "Honestly? I think Mysty was right… about both things."

"Always trust the talking cat," Mysty said smugly from under the table.

Kerilee nodded. "Then we need to get rid of that charm."

Adrienna pushed her plate away. "Take it. Destroy it. And let me know when it's gone."

"You got it." Kerilee wrapped the charm tighter in the napkin. "The demon's playing dirty. We all need to stay sharp."

Adrienna frowned. "How are you going to destroy it?"

Kerilee hesitated. "Good question. I'll ask Margaret. Or Eberhard."

"There's someone better," Mysty said.

Kerilee raised an eyebrow. "Who?"

"You haven't met him yet," Mysty said mysteriously.

"*Him who?*" Adrienna asked.

"I promised not to say until you meet him."

Kerilee groaned. "How am I supposed to meet him if I don't know who he is?"

"He'll come to you," Mysty replied simply.

Kerilee eyed the charm. "So I just... babysit this cursed thing until mystery guy shows up?"

"You could store it in the same material it was hidden in," Mysty suggested.

Adrienna perked up. "The honey frame was plastic."

"Then plastic it is," Mysty declared.

Roger walked over, wiping his hands on a towel. "Need something else?"

"Do you have a plastic container?" Adrienna asked.

"Yeah, we get sauces in them. I'll grab a clean one." Roger headed to the back.

Adrienna stood, glancing around. Only the waitresses remained, wiping down tables, counting tips.

She cleared her throat. "Can I say something?"

Everyone paused. Even Roger peeked out from the kitchen.

Adrienna took a deep breath. "I just... wanted to say I'm sorry. For how I acted."

The waitresses smiled at each other.

"It's okay, Adrienna," one said. "Except for Roger, we're all girls here. We all know any one of us can turn into a monster on the wrong day."

"I rest my case," Mysty said.

A few minutes later, Kerilee dropped the charm into the plastic tub and snapped the lid tight.

Adrienna exhaled. "I actually feel... free."

Kerilee narrowed her eyes. "Makes me wonder... if someone made a cursed charm for *each* of us."

# CHAPTER 9: HE WHO TARRIES

The library smelled of old paper and evening rain.

Kerilee hunched over a cluttered table near the main window, her notebook crowded with scribbled observations and half-translated symbols. Fading golden light slanted through the glass, catching lazy dust motes drifting in the still air.

Mysty sprawled across a stack of books, black fur rumpled, tail flicking.

A low rumble echoed from the vestibule.

Mysty shot upright, fur bristling. Her ears

flattened just before the heavy oak door creaked open.

Kerilee looked up sharply.

John Fisher stepped inside, letting the door swing shut behind him without a sound. In his arms, he carried a box—not cardboard or plastic, but something older: dark wood bound with tarnished brass corners.

Kerilee's spine stiffened, one hand pressing lightly against the open journal in front of her. There was an *otherness* about him—undeniable, unsettling. He hadn't said a word yet, but the air shifted around him like it knew him better than she did. He carried the charge of thunderstorms and secrets too long kept.

John approached slowly and set the box down with a soft thud. Mysty eyed him warily, her gaze narrowing to yellow slits.

"We were just reading about you," Kerilee said, her voice a cautious mix of suspicion and wonder. "Or... someone like you."

John's mouth curved in a ghost of a smile. "Not the first time my name's outlived my face."

He flipped the box's latch and lifted the lid. Inside lay a stack of thin metal plates—etched with flowing, ancient script—and a brittle piece of parchment on top, its edges fraying like old wings. Across it, a title scratched in a shaky hand:

*A Brief History of the Sky Woman and the Demon Mahtantu.*

Kerilee leaned closer. The air around the box felt oddly cooler.

"I've translated what I could," John said, tapping the parchment. "Some plates are readable. Others..." He glanced at Mysty. "Bound. Sealed with a metal you don't mess with lightly."

Kerilee ran her fingers near—but not quite touching—the strange plates. Her heart beat faster, not from fear exactly, but from the heavy, creeping certainty that whatever sat in this box could flip her entire world upside down.

John straightened. His expression softened at the edges, but his eyes stayed shadowed. "Consider it a gift, Kerilee Oberon. You're going to need it."

Before she could answer, a new voice cut through the quiet.

"Well, well. Thought I smelled trouble brewing."

Sheriff Len Embrey pushed the door open wider, boots thudding across the worn wooden floor. His badge caught the fading light as he pulled off his hat, shaking out damp hair. He shut the door with a backward shove of his boot heel.

"I felt like something important was about to get said," he added dryly, though his eyes were sharp as flint.

Mysty let out a soft huff, tail flicking toward the box like a warning flag. "That's a polite way of saying *I always show up late but demand a recap*," she muttered.

Kerilee's chest tightened. Whatever this moment was—it felt like an invisible anchor dropping into her life, and there'd be no pulling it up again.

John reached back into the wooden box and lifted a smaller bundle—metal plates bound tightly with narrow silver bands that glimmered coldly in the fading light.

"These," he said quietly, setting them down with care, "are the plates that are not for reading."

Kerilee leaned forward, curiosity clawing past her better judgment. Mysty, still perched like a sentinel atop the book stack, let out a low, almost imperceptible growl.

"Why not?" Kerilee asked, her voice hushed.

John laid the bundled plates flat on the table. His fingertips hovered over them without touching. "Because these pages hold knowledge meant to be revealed only at the proper time. Sealed by those wiser than we are. Bound against meddling hands."

Mysty's tail twitched. Her ears flattened. "Yeah, well, *someone's always dumb enough to meddle anyway,*" she muttered darkly.

From the shadowed alcove by the genealogy shelves, a figure shifted.

Margaret Fallow.

Kerilee hadn't even noticed her until now —half-hidden in the dim light, pretending to rearrange a stack of old records. But her eyes glinted sharp as glass, tracking every word, every movement at the table.

John noticed, too. His voice dropped to a warning blade. "Some believe hidden power makes them righteous. That secrets are owed to them simply because they seek."

Kerilee pressed her palm flat against the table, steadying herself. "Why are you helping us?" she asked quietly, the question falling heavy between them.

John's gaze flicked briefly toward Margaret,

then back to Kerilee. His expression softened—not weak, but threaded with sorrow. "Because I've seen what happens when no one does."

Sheriff Embrey shifted against the chair, watching John with a mix of trust and unspoken worry.

Kerilee sensed it—a story there, thick with shared history neither man was ready to spill.

Near the shelves, Margaret's fingers tightened around a leather ledger. Her lips moved in a whisper too faint to catch.

Mysty's eyes narrowed. Her tail flicked once, sharp as a whip. "She's already scheming how to crack them open," she muttered under her breath.

John's gaze stayed locked on Kerilee. "The sealed things... they're not just dangerous to the reader. They're dangerous to *everything*."

The weight of his words settled over the room like a shroud.

Kerilee swallowed hard and nodded once. She wouldn't be the fool who opened a door meant to stay shut.

Not if she could help it.

John stood calmly and approached Margaret Fallow who was staring greedily at the place where the sealed plates were.

"Margaret," John said as he blocked her view of the plates.

Margaret shook her head like she was trying to wake up. "John, when did you get here?"

John smiled. "Oh, just a little bit ago. How are you feeling?"

"It's been a strange day."

John nodded. "I can see that. What made it strange, Margaret?"

"I can't, for the life of me, remember what it was."

John became very concerned. "Maybe you should call it a day and go home and rest?"

"You know, John, that's a wonderful idea. Kerilee, could you lock up tonight?"

Kerilee nodded. "Absolutely, Margaret."

Margaret waved goodbye, grabbed her purse from her office, and left.

John returned to his chair. "I think I'm going to take the bound plates with me when I leave."

Mysty put her front paws together as if she were clapping, "Good call, John, and you really handled Margaret well. That was impressive."

"Thank you, Mysty," said John.

Kerilee exhaled slowly and pushed her

notebook closer to the center of the table. "Let's start piecing this together," she said, her voice steadier than she felt.

Mysty leapt lightly down from the books and padded over, tail curling elegantly around her paws. "Best to get the facts straight before Mahtantu pulls any more party tricks."

John nodded, dragging a chair closer, the legs scraping softly across the worn floorboards.

Sheriff Embrey shifted off his perch, standing straighter. He folded his arms across his chest, gaze locked on the table like it held a loaded weapon.

Kerilee flipped to a fresh page, pen poised. "We know Mahtantu isn't... natural. He wasn't born here."

"Banished," John confirmed quietly. "From a world far from this one. Earth was young then. No people yet. Just wilderness."

Kerilee scribbled furiously, the scratch of her pen filling the hush.

Mysty's ears flicked. "And he's allergic to sassafras."

John's mouth quirked faintly. "Repelled by it. Stronger than salt. Stronger than iron, for him."

Kerilee underlined **sassafras** three times.

"That's why Sylvester's trees and root beer matter."

"He feeds on misery," Mysty added, her green eyes gleaming. "Defiance stings him. Betrayal and heartbreak? That's his favorite snack."

Sheriff Embrey's jaw tightened. "And lately, he's getting desperate."

Kerilee looked up. "How do you know?"

Len's voice darkened. "The giant bear wasn't just some freak creature. Mahtantu possessed it. Used it like a puppet. The attack near the sassafras grove... the charm Adrienna found... he's not hiding anymore."

Mysty's tail gave a sharp flick. "Because he knows his clock's ticking."

Kerilee swallowed hard. "Why now? What's changed?"

John's gaze met hers, steady and unreadable. "Maybe the right people finally showed up. Maybe the ones who can stop him are waking up."

The library seemed to shrink around them, heavy with the weight of what wasn't said.

Mysty leaned closer across the table, her voice low and firm. "We're not just solving an old curse, Kerilee. We're fighting something that wants to rewrite its own ending."

Kerilee's pen hovered uselessly above the page. The realization struck her like a bell inside her bones: *this wasn't just history. This was war.*

John shifted, the last rays of sun catching the worn lines on his face. "Sky Woman watches," he murmured. "But she won't intervene unless she's called."

Kerilee leaned in, her hand tightening on her pen. "Called how?"

John's gaze slid to Len.

Silence pressed between them, thick with old knowledge.

Finally, Sheriff Embrey spoke, his voice slow and deliberate. "By her kin. By her blood."

Kerilee's brow furrowed. "You mean... Native blood?"

Len nodded once. "Specifically the Mahican people. My family. My people."

Kerilee watched him closely, sensing layers beneath his words.

"We remember the old songs," Len said, his voice rough with something deeper than years. "Songs that call her name to the stars. My nieces and nephews... they sing them. Only when it matters."

Mysty's tail curled tighter. "Oh, honey. This definitely qualifies as *matters*."

Kerilee's breath caught. She looked at Len, her voice soft but certain. "Could you... take me to them?"

Len studied her, dark eyes unreadable. Then, with a slow, deliberate nod: "When the time's right."

Kerilee didn't argue. Some doors had to open themselves.

John's voice softened, nearly a whisper. "The bridge between old ways and new is thin. Cross it carefully."

Kerilee sat back, notebook momentarily forgotten, feeling the vastness of what she was being asked to believe—and the quiet, ancient power waiting beyond it.

A power that might be their only hope.

Sheriff Embrey's radio crackled against the quiet.

Len unclipped it from his belt, frowning. "Disturbance over on Noord Road. I better check it out."

He nodded once to Kerilee and John, then

strode for the door, boots thudding across the floorboards. A gust of cool air swept in as he left, rattling the papers on the table.

Mysty's ears flicked back. "I don't like this quiet," she muttered darkly. "Too much quiet always means something's sneaking up on you."

Kerilee gathered the scattered papers with a sigh, her mind still spinning from everything she'd just learned. She brushed her hand across the table—then froze.

Something glinted beneath the edge of a folded genealogy chart.

Mysty stiffened. Her yellow eyes narrowed. "Don't touch it."

Kerilee grabbed a pencil and carefully nudged the paper aside.

A charm lay hidden beneath—a small, twisted piece of iron wrapped in blackened twine, oily to the eye despite its dry look.

Unlike Adrienna's honeycomb charm, this one radiated a greasy, wrongness that made Kerilee's stomach knot.

"That doesn't belong here," Mysty said grimly, her voice dropping into a warning growl.

John rose slowly, moving with the caution of a man approaching a bomb. He knelt beside the table, studying the charm without reaching for it.

"It's meant for you," he said, voice flat.

Kerilee's throat tightened. "Me?"

Mysty darted her a sharp glance. "Same foul stink as Adrienna's. But worse. Like someone upped the evil factor just for you."

Kerilee reached under the table for the plastic tub that held Adrienna's cursed charm. She popped the lid and set it open beside the new charm.

John used the edge of a leather-bound ledger to scoop the twisted charm up—careful not to touch it—and dropped it into the container with a metallic clink.

Inside, the two charms pulsed against each other like trapped hornets.

Kerilee flinched back. "I don't like that. That's a bad sound."

Mysty peered over the edge. "Bad? That's a *hell no* sound."

"What is it?" Kerilee whispered.

John's face darkened. "Witchcraft. Crude, but effective."

Mysty's whiskers twitched. "Dimwitty handiwork?"

John shook his head. "They don't have the brains. Someone's guiding them."

Turning to Mysty, Kerilee clucked her tongue. "If it is witchcraft, is it one of the purple witches?"

Mysty shook her head, "No. They work with plants."

"She's right," said John. "We have multiple covens in town, each wearing a different color or shade. This would be the work of one of the darker colors."

Kerilee hugged herself against the cold crawling up her spine. "Can you destroy it?"

John nodded. Then he stood with the plastic container. "I wonder if perhaps Margaret was gifted a charm like these two?"

Kerilee got up. "I could look around Margaret's office if you like?"

John nodded. "Yes. That would explain why Margaret reacted to the plates the way she did."

Following Kerilee to Margaret's office, John found a pair of cheap wooden chopsticks in Margaret's trash can as Kerilee opened each drawer of Margaret's desk. As she opened the third drawer, they both heard something roll and clack against the front of the drawer.

Kerilee pointed. "There it is."

John plucked it out with the chopsticks and dropped it into the plastic container and quickly snapped the lid on it.

"Margaret was being manipulated," John said and returned to his chair. Kerilee followed.

John placed the container in the center of the table and set both hands lightly on the plastic lid.

The air around him seemed to thin, then ripple.

A faint light—neither flame nor electric—glimmered beneath his palms. Kerilee held her breath as the charms shimmered, flickered—

—and then collapsed into fine gray ash.

John exhaled, lifting his hands.

Mysty hopped onto the table and sniffed the tub. She sneezed violently and backed away. "Gone. And *good riddance*."

John brushed his hands off. "It's safe now. Toss the ashes wherever you like."

Kerilee stared at the pile of dust that had once been dark magic.

Her hand tightened around her notebook.

If Mahtantu—or whoever worked through him—was planting cursed charms?

They weren't sending warnings anymore.

They were under attack.

# CHAPTER 10: ROGER'S DREAM

The door to Kerilee's cottage was ajar, and something inside was whispering.

Adrienna frowned as she stepped onto the porch, the weathered boards creaking softly beneath her boots.

The air smelled of autumn leaves and and last night's rain. A soft murmur of voices drifted from the open door, steady and rhythmic, like a lullaby sung by someone trying very hard not to wake the house.

Adrienna knocked once. Then harder. The

sound rang hollow against the old wood.

"Kerilee?" she called, raising her voice.

No answer. But the whispering inside went on, gentle and insistent.

Adrienna hesitated, glancing over her shoulder at the quiet lane behind her. Then she straightened, squared her shoulders, and gripped the doorknob.

"Don't shoot," she called as she pushed the door open. "It's just me. Adrienna."

The air inside was thick with chamomile and something faintly metallic, like old pennies left too long in the rain. The lamps were dimmed beneath gauzy scarves, casting everything in warm, muted tones. It should have felt cozy. It didn't. The whole house seemed to be holding its breath.

On the rug near the hearth, Roger sat cross-legged, head bowed. Kerilee knelt behind him, her hands lightly cupping his temples. Her lips moved steadily, whispering words Adrienna couldn't quite catch—half-heard syllables that vibrated at the edge of understanding.

Mysty stretched across Kerilee's lap, yellow eyes half-lidded, paws twitching like she was chasing something only she could see.

Adrienna blinked, disoriented.

She stepped forward—

—and the moment her foot crossed the threshold, a wave of dizziness slammed into her.

She stumbled, reaching for the back of a chair—only for her hand to swipe through empty air.

The floor tilted beneath her. The walls stretched and blurred. A low hum filled her ears, like bees trapped inside a jar.

"Kerilee?" Adrienna tried to shout, but her voice came out thin and distant, swallowed by the shimmering air.

Her heartbeat thundered in her chest.

She squeezed her eyes shut.

Then—

The sensation of falling. Weightless. Spinning.

A flash of blinding white—

—and the soft, grassy earth caught her like a net.

Birdsong. Sunlight. The sweet scent of honeysuckle.

Adrienna sat up, blinking against the sudden brightness.

Ahead, Kerilee and Mysty walked beside Roger

toward a bubbling stream. Beyond it stood a tall white country house with pale blue shutters, ivy crawling up its stone façade.

In the yard, a girl swung beneath an old oak tree, her laughter bright and clear, ringing like wind chimes.

Adrienna watched as the girl leapt from the swing, landing lightly. She ran straight to Roger.

He didn't stop her.

He didn't hesitate.

She threw her arms around him, and he let her.

Adrienna's stomach flipped.

She marched forward, fists clenched. "What is going on?"

Roger stammered. "Adrienna, I—"

Kerilee laid a hand on his shoulder. "Let me."

She turned to Adrienna. "This isn't Roger's choice. When he dreams, he ends up here. Mahtantu sends him."

Roger nodded, shame reddening his cheeks. "To make me lose you."

The girl took a step back. "Are you making me the villain here?"

"Joanna," Roger said, reaching for her hand. "I don't want to hurt you, either."

From the porch, two men approached. One was lean and silver-haired. The other...

"John Fisher," Kerilee murmured.

John tilted his head. "Do I know you?"

"Not yet," she said. "We're from your future."

John blinked. "Well, that's unsettling."

"It's like this," Kerilee rushed, "a demon cursed Roger, who's dating Adrienna, but the curse sends Roger to Joanna when he sleeps, so they came to me to help them."

John nodded thoughtfully. "Ah. That explains... more than I wanted it to."

Joanna's father, Victor, looked bewildered. "This entire conversation is lost on me."

Roger turned to Joanna. "In your time, I haven't even been born yet. When I sleep in my time, I wake here. And when I sleep here, I wake back home."

Joanna's eyes widened. Then she said, slowly, "So I'm not stealing you from anyone... because she doesn't even exist yet."

Roger flinched, guilt twisting his features. He looked like a man trapped inside a collapsing house with no doors, no windows. His hands curled into helpless fists.

"I never wanted this," he rasped. "Not to hurt

her. Not to hurt anyone."

Adrienna's mouth opened—then clamped shut.

Joanna blanched. "I didn't mean it like that."

Adrienna bit her bottom lip, fury flashing hot through her. She made a fist and lunged forward—

Kerilee stepped between them instantly, catching Adrienna's wrist mid-swing. "Whoa, whoa, whoa!" she said firmly, locking eyes with her.

Adrienna struggled for half a second, furious tears in her eyes, but Kerilee's grip held steady—strong enough to stop her, gentle enough not to hurt.

Mysty hopped down from Kerilee's shoulder, padding around them with a slow, knowing circle. Her yellow eyes gleamed with dry amusement.

"Punching ghosts doesn't fix living problems," Mysty said casually. "But ten out of ten for commitment."

Adrienna's breathing slowed. Her fist uncurled under Kerilee's calm gaze.

"We need to focus," Kerilee said softly. "This all ends when Mahtantu ends."

Adrienna pulled her arm back, tension still coiled under her skin.

Victor threw up his hands. "Well, if there's going to be violence! I'm going inside." He stalked toward the house.

John pulled Joanna aside. "Come. We need to talk."

The air shimmered.

Kerilee grabbed Roger's hand—and Adrienna's without thinking.

The world snapped.

They awoke with a gasp on Kerilee's cottage floor. The floorboards were cold beneath them. Mysty blinked rapidly, fur fluffed up.

Roger lay still, lips parted in a soft whisper: "Joanna…"

Adrienna turned away.

Kerilee helped her sit up. "I'm sorry you had to see that. We had planned to do this without telling you just so this kind of thing wouldn't happen."

Adrienna exhaled shakily. "I get it. I just wish I didn't."

Mysty padded to the window—and froze.

Her yellow eyes narrowed. Her tail fluffed. "Something's not right."

Kerilee steadied Adrienna with a hand on her arm. Roger still lay slumped on the rug, looking heartbreakingly young, fragile. He murmured again, so softly it barely reached them: "Joanna…"

Adrienna stiffened.

For a moment, she looked ready to bolt, fists clenched at her sides. Then her shoulders sagged. She scrubbed a hand down her face, her voice low and raw.

"I can't even be mad at him," she muttered. "That's what Mahtantu wants."

Kerilee crouched beside Roger, brushing hair gently back from his forehead. "He's fighting it," she said quietly. "Harder than you know."

Adrienna nodded, but the storm hadn't left her eyes.

At the window, Mysty stood statue-still, tail bristled, her gaze locked on the fading twilight. "We brought back more than memories," she said grimly.

Kerilee's head snapped up. "What do you mean?"

Before Mysty could answer, a sharp *crack*

echoed from outside—like brittle branches breaking, but deeper, heavier.

Adrienna jumped. "What was that?"

Kerilee shot to her feet, tension rippling through her. She crossed to Mysty, standing shoulder to shoulder with the cat at the window.

"What do you see?" she whispered.

Mysty's tail lashed once. Twice. "Nothing I like."

Roger stirred on the floor, groaning faintly.

Kerilee didn't dare look away from the window. She could feel it now, crawling across her skin: *a ripple of wrongness, riding the night air.*

Something—or someone—had felt the dream shift.

And it was coming.

Mysty's voice lowered, dry and dark: "Oh goodie. Company."

A deep, low rumble rolled beneath the floorboards.

Kerilee pressed her hand against the wall, feeling it vibrate like a distant drum.

Mysty stiffened at the window, tail puffed full. Her ears flattened. "Oh, that's... big."

Outside, something groaned—a sound like timber under strain, mixed with grinding stone.

Adrienna's breath hitched. "Is it Mahtantu?"

"No," Mysty said shortly, eyes locked on the dark beyond the glass. "Worse."

Roger stirred on the rug, blinking groggily. "What's happening?"

The rumble deepened. Across the street, a lamppost flickered wildly, then snapped with a sharp *crack,* sparks raining down as the ground trembled again.

Adrienna ran to Kerilee's side. "Kerilee—look—"

From the corner of her window, Kerilee saw it:

A massive shape moving between the rooftops, too large to fully comprehend, like a piece of the mountain itself had risen and walked. In the dim light, the outline of a colossal leg lifted, paused, then swung gently forward—stepping delicately *over* a row of houses as if they were toys in its path.

The ground shuddered under its weight. Shingles rattled down a rooftop nearby.

"Holy—" Adrienna clamped both hands over her mouth.

Roger crawled weakly toward the window, peering up. "That's not real."

"It's real," Mysty said flatly. "Stone Giant."

"*Stone giant?*" Kerilee echoed, heart hammering.

Mysty flicked her tail once, calm settling back into her voice. "Ge-no-sgwa is its name. Old as the hills. Strong as the hills. Literally."

The giant's shadow swept across the cottage as its foot came down somewhere beyond the next block, shaking the shelves and rattling picture frames.

"Is it coming back?" Adrienna whispered.

"Nah," Mysty said. "Not interested in you." She turned from the window, hopping lightly to the floor. "Probably following an old path. Or maybe chasing something nastier than you."

Kerilee swallowed hard. "I thought it was going to crush us."

"Could've," Mysty agreed breezily. "Didn't."

Adrienna slumped against the wall. "That's your reassurance?"

"You're still standing, aren't you?" Mysty's whiskers twitched. "Trust me: if he wanted us? We'd be a *Kerilee-and-friends pancake special.*"

Roger let out a shaky breath. "I've never... I didn't even know something like that existed."

Mysty smirked. "You've lived here your whole life, never met one. And you're still alive. See?

Most folks never do."

Kerilee kept staring out the window, watching the last glimpse of the giant's silhouette disappear over the trees like a vanishing hill. "Why didn't it stop here?"

"Not its business," Mysty said. "Stone Giants follow their own rules. Hunt their own prey. Leave their own tracks. We're just... scenery."

Adrienna's voice rose nervously. "Are there *worse* things out there?"

Mysty's grin widened. "Oh, absolutely." She flicked an ear toward the door. "But worse things usually knock. He just walks on by."

Kerilee let out a small, incredulous laugh, rubbing her eyes. "You're terrible at comfort."

"I keep you alive," Mysty replied smugly. "Comfort's optional."

Another faint tremor rolled beneath them. Then quiet.

The shelves stopped shaking. The last lamppost flickered steady again.

Kerilee exhaled slowly, letting the tension drain from her shoulders. "Is it... gone?"

"Gone," Mysty confirmed, settling into a loaf on the rug. "And if we're lucky? He won't be back anytime soon."

Adrienna finally pulled herself away from the wall. "If that's *lucky*, I'm terrified to know what unlucky looks like."

Mysty stretched, closing her eyes. "Unlucky doesn't knock either."

Kerilee stared once more into the night, the deep hum of the village returning to stillness around them.

Whatever walked tonight had passed them by.

But something told her—it was not out of mercy.

# CHAPTER 11: CATALOGING GARDEN ROCK

The wind shifted sharply as the Sheriff, Kerilee, and Mysty reached the ridge above Kaaterskill Falls.

Mysty hesitated, ears flattening briefly. "Still feels like something's trailing us," she muttered. Kerilee tightened her grip on the satchel but said nothing. Whatever it was, it hadn't shown itself yet. Kerilee Oberon paused on the trail, her breath catching in her throat. The air was colder here—thinner somehow, sharper. Beneath the crashing rush of water, another sound pressed close— A low, constant hum, like distant voices trying not to be forgotten.

Sheriff Embrey was already a few steps ahead, his hand resting on the hilt of his holstered weapon. Mysty padded beside Kerilee, ears angled back.

"This place doesn't just remember," Mysty murmured. "It mourns."

Kerilee adjusted the strap of her satchel and followed. The trail narrowed, winding between blackened trunks that loomed overhead. As they ascended the final bend, the woods opened abruptly into a clearing.

Garden Rock.

Locals also called it the Devil's Grasp, and now Kerilee understood why.

At its center lay a pond, unnaturally still, its surface mirrored with lily pads and drifting silver water snakes that moved as if directed by an unseen rhythm. Around it stood the trees —twisted, humanoid forms with trunks shaped like legs and arms raised toward the sky. Some had gnarled "hands" outstretched in silent screams. Each one bore a gourd in place of a head.

Kerilee stepped into the clearing and stopped. "They look… like people."

"They were," Mysty said.

Sheriff Embrey kept his distance from the pond. "I grew up hearing not to come here. My uncle said the pond was cursed. That things crawled out of it at night."

Kerilee knelt beside one of the trees and ran a finger down its bark. It was rough, like scar tissue. The arms were crooked but unmistakably human in proportion. She flipped open her notebook and began sketching.

"Seventeen... eighteen..." she counted softly. "Nineteen."

Kerilee flipped through her notes, fingers trailing down the list she'd been piecing together from town records, Rip's scribbled margins, and old newspaper clippings.

"These aren't just trees," she murmured. "They really are people. Every one of them. Family."

She stepped toward a thick, broad-shouldered trunk near the pond's edge. Its arms reached upward in defiance. "Ezra Van Winkle," she said. "Disappeared 1794. Left for Albany, never came back. His fiancée waited years. His name's still carved on a headstone in the old churchyard."

Sheriff Embrey's brow furrowed. "Folks knew he vanished, but no one came looking here?"

Kerilee's lips pressed tight. "They looked. They

just didn't know what they were seeing."

She pointed to a smaller tree, its hands clasped together like in prayer. "Sarah Van Winkle. Vanished 1823. Tip's great-great-aunt. The family said she drowned, but Rip's notes said she disappeared by Garden Rock. They found her bonnet floating in the pond."

Sheriff Embrey shook his head slowly. "Every one of 'em... right under our noses."

Kerilee moved to another, crooked and leaning sharply as if collapsed into itself. "Henry Van Winkle. 1881. Tip's grandfather's cousin. He used to run whiskey stills in the hills. Thought he was clever. Mahtantu didn't appreciate that."

Mysty flicked her tail. "Not big on small business owners, huh."

Kerilee managed a brief smile, then sobered as she approached the slender tree with long hair carved down the back.

"This one..." She paused, swallowing. "LeeAnn."

Sheriff Embrey's breath caught. "Tip's wife."

Kerilee nodded. "Disappeared 2005. Folks thought she left her family. But Tip... he never gave up."

Mysty padded closer, her voice quiet. "He was

right not to."

Kerilee ran her hand gently along the bark, feeling the cold. "And Sylvester... he's been painting her face for Tip all these years. Every year. Trying to hold onto her."

Sheriff Embrey's voice was grim. "And she was here the whole time."

Kerilee looked slowly around the clearing, the trees no longer just twisted wood—but faces, names, stories.

"This isn't random," she murmured. "It's deliberate. Mahtantu doesn't just curse the Van Winkles. He curses what they love."

Mysty's gaze turned toward the pond. "And those gourds? Might not just be decoration. They're vessels. For their souls. Or for the curse."

Kerilee pulled a folded page from her satchel, scanning her notes. "Washington Irving wrote about a hunter who once came to Garden Rock. Said he found gourds tucked into tree branches. He stole one, dropped it, there was an explosion and a flood poured out of the earth—turned into Kaaterskill Falls."

She looked slowly at the gourd-headed trees. "But if these gourds are... their heads... then the person in that tree..."

Mysty's tail lashed. "Well, he's definitely out of his gourd."

Sheriff Embrey's jaw tightened. "I don't think it was a vessel for a soul or the curse."

Kerilee's voice went quiet. "Let's hope we don't have to find out."

A ripple broke the pond's glassy surface.

All three turned.

A silver snake leapt—striking toward Kerilee.

Sheriff Embrey drew and fired without hesitation. The bullet cracked the surface, and the creature vanished in a swirl of ripples.

But more ripples followed.

Another silver snake slithered up from the pond's edge, and a second burst from the lily pads across the water, both angling toward them with frightening speed.

Sheriff Embrey fired twice more, splashing the surface with sharp cracks. The silver shapes recoiled and disappeared beneath the water.

Kerilee stumbled back, breathing hard.

"Let's not stay until sundown," Mysty said.

Kerilee gathered her notes. "We've seen enough."

As they turned to leave, a gust of wind tore through the clearing, snatching pages from

Kerilee's notebook. She caught most—but one sketch, the tree with the long hair, drifted.

It hit a tree trunk.

A gnarled branch unfurled slowly.

And pulled the sketch inward.

Sheriff Embrey stepped forward, raising his gun again. "What the hell—"

"Don't," Kerilee said quickly. "It… that's a tree we can save one day."

They hurried back down the trail. As the trees closed in behind them, a low hum returned—quieter now, but full of sorrow.

Kerilee didn't speak until the falls were behind them.

"She's not lost. She's still here."

Mysty nodded. "Then we need more than love to bring her back."

"What do we need?"

Mysty's eyes flashed. "The one thing Mahtantu can't stand."

"The brew that cures the curse?"

"Which is also like poison for him."

# CHAPTER 12: THE RECIPE TAKEN TO A PROFESSIONAL

Arnoldus De Groot

The library was quiet—too quiet.

Kerilee moved quickly to the reading table and unfurled the sketches she'd made at Garden Rock. The tree with the long hair. The one with the missing gourd. The carved arms stretched to the sky, as if pleading.

"Kerilee flipped to the back of the real Rip Van Winkle's journal that Margaret kept locked away. The ink turned frantic, half-sentences running together. One phrase repeated: I was commanded not to write it down."

She traced it with her finger.

A folded scrap caught her attention near the spine. A sketch. Crude, but unmistakable. Garden Rock—labeled in Rip's hand as The Demon's Grasp.

Beneath it, another line: All with gourds are my kin.

Kerilee's breath hitched. "He knew."

Mysty padded closer. "Flip the page. There's more."

Kerilee turned the page under the lamplight.

Her gaze drifted to the sketch she'd made of the long haired tree. LeeAnn. Even here, surrounded by books and lamplight, the thought of Tip's wife trapped in wood and gourd made her chest ache. "We're coming for you," she whispered, though no one was there to hear.

Faint ink shimmered into view. Mysty's eyes narrowed. "The flute of many pieces shall call the Pukwudgies only when the time is right."

Kerilee stared at her. "That wasn't there

before."

"You couldn't see it," Mysty said. "But I could."

Kerilee rummaged through the journal and found a tucked-in side note. The paper was rough and yellowed, written in smaller, tighter script: Once, they were friends of man, then evil ones turned them away from us.

She whispered, "Does he mean these Pukwudgies?"

Mysty nodded slowly. "Yeah. Also, Eberhard once told me the land had a knowledge of when things would happen. Said that knowledge would reveal itself in its own time and that would include the Pukwudgies."

Kerilee's stomach sank. "Then he's the only one who would know the when and the where."

She looked toward the nearest window. Night had entirely fallen. "We have to find him. Now."

"Can we walk this time?" Mysty whined. "I hate getting car sick."

"Sure, it isn't far"

Tip Van Winkle's house was dark except for a single lamp in the front room. The door creaked as Kerilee pushed it open.

"Tip?" she called.

"Back here," a woman's voice echoed.

She found Karyn seated alone at the kitchen table, reading a note. The Sheriff leaned against a doorframe.

"Do you know where Eberhard is?"

Karyn blinked. "Tip left me a note. Said Eberhard was taking him on a trip." She handed her a folded square of parchment. On it, in tiny, elegant script: Pack light. Your needs shall be provided.

Kerilee frowned. "Well, finding Eberhard soon doesn't sound promising."

"He was supposed to come by the diner two days ago. Didn't. And that little guy's never late. Not once."

"That's why we're here," said the Sheriff.

Mysty paced the floor. "He's never absent without reason."

Kerilee turned to the Sheriff, "When did you last see him?"

Len Embrey answered, "The morning before we climbed to Garden Rock. He was waiting by the truck. Gave me that look—you know the one that feels like he knows what dream you had last night."

Kerilee let out a long breath. "Well, if neither of them are here, then we'll just head back to the library."

As Kerilee and Mysty made their way down the winding country road towards town, the night grew unnaturally still. No crickets. No breeze. Just a heavy, listening silence.

They passed an ancient oak tree, gnarled and sprawling over the path. Mysty halted, ears flattening.

"The land knows," she whispered.

Kerilee slowed, her instincts prickling. She glanced up at the oak—and for a heartbeat, she thought she saw its branches shift ever so slightly, not with the wind, but with intent.

"Not just remembering anymore," Mysty said. "The land's choosing who it will protect--and who it will forsake."

Kerilee tightened her hold on Rip's journal and kept walking, heart pounding.

Back at the library, Kerilee spread the recipe for the Brew that Eberhard gave Rip Van Winkle.

"The recipe mentions brewing using outdated methods. Probably because they did not know any better method at the time."

Mysty circled the table. "We have the recipe. However, we need someone who can understand how to brew it with what we have available in our day."

"Do you know anyone?" Kerilee asked.

Mysty nodded. "Of course. Adrienne's dad used to work for a major soft drink manufacturer. He is exactly the kind of person we need."

"Where does he live?"

"In his mansion on the edge of town."

"A mansion," said Kerilee.

"Yeah, a nice one with a lab."

"Would he be up at this hour?"

"For a recipe like this one," Mysty said with a grin, "he wouldn't care what time of night you woke him up!"

Later that evening, Kerilee and Mysty arrived at the De Groot mansion. The winding drive led them through iron gates and manicured hedges. The house loomed like an old European villa, complete with ivy-covered stone and ornate lanterns flickering with gaslight.

Arnoldus greeted them in a long smoking

jacket and velvet slippers. "Ah, the infamous librarian and her feline critic," he said with a bow.

"I brought the recipe," Kerilee said without delay.

"And I brought opinions," Mysty added.

Arnoldus led them down a long hallway lined with portraits of stern-faced ancestors and into a laboratory that looked part science lab, part apothecary. Along the way, he paused beside a glass display case, tapping it lightly. Inside lay a curious object: a small crystal flute, gleaming under the dim lights.

"A rare piece," Arnoldus said. "Some say Mozart himself immortalized its legend as a magic flute—based on true stories he heard in his youth. Perhaps not all fairy tales are fiction."

Kerilee leaned closer, drawn by the intricate etchings swirling along its surface. Mysty sniffed delicately. "Old magic," she murmured. Glass jars in the lab lined the walls, each labeled in cursive —Sassafras Root, Gentian Bark, Smoked Vanilla Pod.

Kerilee handed him Rip's recipe. Arnoldus adjusted his glasses and read.

"Sassafras, spicy roots, night berries..." he

trailed off. "This is a challenging brew. Potent. Dangerous, even, if mishandled. But effective."

"We're short a few ingredients," Kerilee added.

He nodded slowly. "Oh, I can see that. That's no problem. We both know of a little man who can help us find those."

"Eberhard," Mysty said. "But, no one knows where he is."

Arnoldus turned the page. "Spicy roots... yes. I have access to some of these. But one —'Scorchroot'—grows only in limestone caves. You'll need someone small and clever to retrieve it."

Mysty smirked knowingly. "Of course. Pukwudgies."

"So, you know of them—incredible creatures. Someone will need to find all the hidden pieces of that wooden flute that calls them," Arnoldus said without looking up. "Obviously, not the one in my collection."

Kerilee blinked. "You know about the Pukwudgie flute?"

Arnoldus finally met her gaze. "Of course! This area is filled with wonder for those with eyes to see it. Your eyes, like my daughter's eyes, are just waking up to it all."

They left shortly after, the recipe marked with Arnoldus's precise notes.

As they reached the front door, he called out, "Be cautious. Brewing is a science. But what you're doing? That's something older. It's alchemy. And alchemy always demands a price."

Turning back, Kerilee asked, "What price is demanded?"

Looking at his shoes, Arnoldus worked up the nerve to answer. "I did say this recipe could be dangerous if mishandled. There is something called Sacred Geometry. This recipe will require a close cousin that we call Sacred Chemistry. The elements required will be taken from those that powers above us judge they should come from. It is usually a sacrifice of some sort. This recipe is not forbidden, so I am pretty sure no one's life will be required. However, pain, a hard struggle, loss of freedom, or severe injury may befall several of us. We won't know who will have to endure such things until we all push ourselves through to the end."

Mysty was suddenly sitting before Kerilee looking up at her. "It is to fulfill the laws of eternal balance. The price is always balance between the opposing forces. Mortals usually never see how out of balance things get, and so they think the result is unfair. Yet, if you were a higher being, you would see how it all plays out properly."

Suddenly, Kerilee's vision was pulled deeper—not just flashes, but a vivid, living sequence.

First, she saw Revolution-era soldiers shouting harsh orders. A massive cauldron boiled furiously, thick black smoke curling up to the sky. People in brown robes rushed through a crumbling stone tunnel, carrying bundles of scrolls and ancient artifacts. Their faces were lined with fear—but also fierce determination. Among them, Kerilee thought she glimpsed a flash of familiar hair—Adrienna? Joanna? Roger?

As if a curtain had fallen and risen again on a different stage, the vision shifted.

The underground city opened before her: vast walls and ceilings glittering with embedded crystals. Winding streets twisted between stone cottages and fields of golden mushrooms. Legendary small folk from every corner of the world mingled there—leprechauns, brownies, gnomes, kobolds, and aluxob.

In the center of a great hall, Tip Van Winkle and Eberhard sat together at a carved table laden with strange fruits and sparkling drinks. They laughed, clinking stone goblets, surrounded by music and warmth.

A deep, reassuring joy flooded through her—a sense of belonging, of ancient promises still kept.

Mysty's voice pierced through the haze. "Kerilee, are you alright? Kerilee!"

Kerilee staggered, breathing hard. "Whoa, that hit me hard," she whispered.

Mysty didn't blink. "That happens when the veil peels itself back."

Kerilee wiped the sweat from her brow, already dreading what might follow.

# CHAPTER 13: NIECES AND NEPHEWS

The morning sun poured thick and golden through the high windows of the honey windmill, wrapping everything it touched in the sticky warmth of late autumn. Adrienna de Groot wiped the back of her wrist across her forehead, leaving a smear of honey along her temple she didn't notice. Bottles clinked gently as she lined them up on the worn oak counter, each filled with syrupy amber treasure fresh from the comb.

She was humming to herself—something mindless, just to fill the empty space—when the front door creaked open.

Adrienna turned, brushing her sticky gloves against her apron. Her heart gave a little jump when she saw who stepped inside.

Kerilee entered first, looking more serious than usual, followed by Sheriff Embrey. But it wasn't just the two of them. Behind them came five teenagers, each carrying a mix of eagerness and nerves, their boots dusty from the morning trail.

Adrienna tugged off one glove with her teeth. "Uh… morning?"

"Morning," the Sheriff said, his voice oddly formal. "Adrienna, these are my family. Ruth, Phil, Jonameed, Lucy, and Marcus. We thought it best to meet here—neutral ground. And besides, no one would expect the fight to begin in a windmill."

The teens fanned out slightly, as if unsure how close to stand. One by one, they stepped forward.

Ruth, tall and poised, gave a slight bow. "It's an honor to meet you, ma'am."

Phil, lean and restless, offered a crooked smile. "We've heard a lot about you."

Jonameed, stocky and serious, nodded deeply without speaking.

Lucy, a wild-haired girl with eyes too sharp for her age, grinned. "Finally meeting the honey lady!"

And Marcus, the youngest, who couldn't have been more than fourteen, shuffled his feet and muttered, "Hi," before staring intently at the floor.

Adrienna blinked at them, feeling oddly displaced in her own shop. She wiped her hands on her apron again, unsure what to say. These were kids. Teenagers. How in the world were they supposed to help fight a centuries-old demon curse?

Still, there was something different about the way they held themselves. Something that prickled at her instincts.

"They're here to help," Kerilee said gently, as if reading Adrienna's doubts before she could voice them.

Adrienna managed a shaky smile. "Right. Well... welcome to the Windmill."

Adrienna led them inside, peeling her gloves off fully now, trying not to look as overwhelmed as she felt. The parlor of the windmill smelled of beeswax and clover, the old wooden beams arching overhead like ribs in a giant, ancient beast.

Kerilee motioned for everyone to gather around the low circular table near the stone fireplace. The teens sat without needing to be told where—each one seeming to instinctively find a spot, as if they'd done this a hundred times before.

Adrienna perched awkwardly on the edge of a rocking chair, feeling about three steps behind whatever strange dance was unfolding.

Ruth, the tall one, folded her hands neatly in her lap. "We're part of the old ways," she said, her voice calm but carrying weight. "Sheriff Embrey taught us. And others before him."

"We're not here by accident," added Phil, tapping the table absently with his fingertips.

One by one, they began to speak.

Ruth first: "I study the stories of Sky Woman—the rites, the songs, the star patterns. I know when the signs appear, what they mean."

Jonameed leaned forward, his dark eyes steady. "I know Mahtantu's forms. How he hides. How he hunts. My grandfather told me all the tales that weren't written down."

Lucy, bouncing a little where she sat, pulled a thin silver whistle from her jacket pocket. "I'm the keeper of Pukwudgie lore. And... I have this." She twirled the whistle between her fingers, catching the light. "It's old. Older than the town. Meant for emergencies. Really big ones."

Phil shrugged modestly. "I remember places. Old ones. Where the ground's thin. Where lines of power run like rivers underground. I can walk where history forgot."

And Marcus, small but steady, spoke last: "I set traps. Wards. I know how to fight things you can't punch."

Adrienna stared at them, her mouth slightly open.

"You've all been preparing for this?" she managed.

Ruth nodded once. "Our whole lives. We just needed the sign that it was time."

A silence settled over the room, thick and humming with things unsaid. Adrienna could feel it pressing against her skin—the gravity of it.

These kids weren't playing at being heroes. They had been forged for it.

Kerilee unfolded a sheet of paper from her satchel and spread it on the table between them.

It was a rough sketch—the mural they had found in Tip Van Winkle's hidden basement, drawn in quick but careful strokes.

"We found this under Tip's farmhouse," Kerilee said, tapping the center of the page. "A mural. Symbols, vines, gourds... and something that looked like a flute in pieces."

The teens leaned in, studying it closely. Ruth's brow furrowed as she traced a finger along one of the painted gourd shapes.

"Sacred vessels," Ruth murmured. "Bound to the spirit. If sacred names were painted near them, they might be clues."

"To reversing the curse?" Adrienna asked.

Ruth nodded gravely. "Or at least guiding restoration. Sometimes the old magic isn't about breaking things—it's about mending what was torn."

Phil tapped the lower edge of the sketch. "This mural sits on a ley point. I'd bet anything. If I can map the surrounding lines near Garden Rock, I might find the strongest place to summon help."

Lucy hugged the silver whistle to her chest. "And if we need the Pukwudgies... I'll be ready." Her voice dropped to a whisper.

Adrienna shivered slightly at the way

Lucy said it—matter-of-fact, no drama. These weren't ordinary teenagers. They carried ancient burdens in their blood.

Kerilee rolled up the sketch carefully. "We'll need all of you if we're going to do this right."

For the first time that morning, Adrienna felt a twinge of hope.

Not because the danger was any less—but because maybe, just maybe, they weren't facing it alone.

Adrienna sat back in her chair, watching the teens talk among themselves, their heads bent close together like a council of young sages.

Lucy was showing Marcus how the whistle's different sections could twist apart; Phil and Jonameed were tracing invisible paths across the wooden table, murmuring about landmarks and ancient trails.

They carried themselves with a sureness that Adrienna couldn't help but envy.

They know exactly who they are, she thought, the realization landing heavy in her chest. Why don't I?

Here she was—older than all of them, supposedly one of the "adults" leading this effort —and yet half the time she still felt like she was scrambling just to keep up, trying to patch over fears she didn't want to name.

She glanced down at her own hands, callused from years of tending bees and boiling honey, and wondered if it was enough.

A soft voice interrupted her thoughts.

"You've done more than you think," Kerilee said, slipping into the seat beside her. Her eyes were kind but steady. "You helped us believe. That counts for more than you know. Remember when I first visited your windmill? You had to convince me to get involved, and you succeeded."

Adrienna looked at her, emotions knotting her throat. She hadn't realized how much she'd needed to hear it until the words were spoken.

Before she could answer, Mysty, lounging like a queen across a windowsill, flicked her tail lazily and added, "Not everyone gets handed a silver whistle and a map. Some of us have to build the road while we walk it."

Adrienna gave a shaky laugh. Trust Mysty to sum up the impossible in one line.

She squared her shoulders slightly.

Maybe she wasn't forged for this like the others.

Maybe she was still figuring herself out.

But she was here. And she wasn't running.

The circle around the table tightened instinctively, as if everyone sensed the conversation had reached a threshold.

Sheriff Embrey shifted his weight, the leather of his belt creaking softly. His voice came low and firm.

"We'll visit the Van Winkle house next. Tip's place. That basement... there's more waiting there than what you found."

Kerilee frowned. "Like what? Another mural? Or something worse?"

The Sheriff didn't answer. "We'll see."

Adrienna glanced at the teens. None of them flinched. They leaned forward instead, their eagerness tempered by something sharper—readiness.

"But listen to me," the Sheriff continued. "If Mahtantu's paying attention—and he will be, now—he's not going to like us poking around. Especially not all together."

Ruth nodded solemnly. "The old stories say he fears unity more than weapons."

Lucy tightened her grip around the whistle. She looked smaller suddenly, despite her fierce spirit.

"Then I hope the Pukwudgies are listening," she said.

The room felt heavier, as if the very walls were holding their breath.

Kerilee tucked the mural sketch safely back into her satchel and rose. "We move carefully. No bravado. No unnecessary risks."

Adrienna stood too, brushing invisible dust from her apron. Her hands were steady now. Her decision made.

"We're ready," she said simply.

Mysty leapt from the windowsill, tail high, yellow eyes bright.

"Then what are we waiting for?"

They filed toward the door, the weight of history at their backs—and something unseen waiting ahead.

Adrienna paused just before stepping outside.

A sudden ripple ran up her spine, sharp and icy, like someone had brushed her soul with a cold hand.

Then it came.

A scream—not human, not animal, but something deeper and more primal—ripped through the air.

Adrienna's knees nearly buckled at the sound.

Ruth whispered, "That's not his hunting cry."

The ground trembled under their feet. Bottles rattled violently on the shelves. Bees from the hives outside erupted into frenzied, panicked flight.

Adrienna froze, breath locked in her chest.

The scream rolled again, farther now, echoing over the hills like thunder dragging itself across stone.

Mysty's fur stood on end. Her voice was a taut wire.

"That wasn't just rage," she said. "That was panic."

And for the first time, Adrienna realized —Mahtantu wasn't simply angry they were moving against him.

He was afraid.

# CHAPTER 14: THE RETURN TO RIP'S BASEMENT

The sky hung heavy over Molenhaven, low and bruised with unfallen rain. Humidity pressed against the earth like a wet cloth, thickening the air and leaving the skin clammy. It was late morning, but the light felt dim and uncertain as Kerilee Oberon followed Sheriff Embrey up the rutted gravel driveway to the Van Winkle home.

Mysty trotted at her heels, fur sleek despite the damp. Adrienna walked beside her, arms crossed tightly over her chest as if to ward off the prickling unease that clung to them all.

Behind them, the five teens moved in a loose cluster—Ruth, Phil, Jonameed, Lucy, and Marcus—each carrying packs slung over their shoulders and expressions carved from wary determination.

Sylvester Van Winkle met them at the sagging front porch, his eyes hollow with a kind of permanent grief that no time seemed able to mend.

"You're here," he said simply, voice rough. "Good. Come in. It's ready."

Inside, the house smelled of old wood, drying herbs, and the faint, sharp tang of gun oil. The floors creaked under their boots. Without preamble, Tip led them down a narrow hallway, his limp more pronounced today, and pulled aside a heavy wool rug to reveal the trapdoor.

Kerilee glanced once at Adrienna, who gave a small nod. Together, they descended the steep wooden stairs into the basement.

The air below was cool and musty, lit by the flickering glow of oil lanterns hung at intervals along the walls. It wasn't a large room, but it felt vast somehow—every surface heavy with history and sorrow.

And wrapping the stone walls like a living, breathing thing was Sylvester's mural.

Even Kerilee, who had seen it before, felt the impact anew. It wasn't just paint. It was spirit and memory, poured into every curve and color.

Figures twined among thick vines: sleeping men, fierce beasts with twisted horns, glowing gourds hanging from tree trunks. And woven through it all, the unmistakable presence of a blue, burning shadow—sometimes hidden, sometimes looming.

Ruth stood back, her breath catching audibly.

"It's like a painted spell," she whispered. "The good kind."

Jonameed nodded slowly beside her, reverence softening the set of his broad shoulders.

Sylvester leaned against the wall, catching his breath. "It was here before my time. Before my father's, too. Maybe even back to Rip himself."

Rip Van Winkle had left more behind than just a story.

And somewhere in this room, the answers they needed were waiting to be found.

Kerilee moved slowly along the wall, fingertips trailing lightly over the ancient mural.

Some symbols were obvious: sleeping men caught mid-dream, beasts mid-roar. But others were stranger—spirals, crossed vines, stars stitched among twisted trees.

Ruth stepped closer, her hands tucked behind her back like a respectful scholar at an ancient tomb.

"These marks," she said, pointing to a series of spiral knots etched near the center, "they're protective wards. Old ones. Tied to the sleeping world. That means... someone already tried to shield this place."

Jonameed knelt by a tangled set of roots painted near the floor, brushing dust from the crevices.

"And these," he added, tapping a pattern of claw marks barely visible beneath a faded vine, "are warnings. Mahtantu's passage—burning, twisting, devouring."

Sylvester shifted uneasily against the far wall. "Those parts of the mural... they were already here when I was a boy. They feel older. Like some ancestor painted them before memory."

Kerilee paused at a curling spiral tucked low near the floor, half-obscured by soot. Something about it prickled at the edge of her mind. She crouched, brushing the stone clean with the sleeve of her jacket.

The spiral wasn't just a symbol.

It was a mechanism.

She pressed her palm against its center.

A deep, muffled click echoed through the basement. The mural shivered at the edges—then a small panel, no larger than a breadbox, popped open with a dusty hiss from the wall.

Everyone leaned forward, the lantern light catching on something inside.

Kerilee reached in carefully and withdrew two items:

—An ancient scroll, wrapped tightly in cracked waxed cloth.

—And a heavy envelope, sealed with a perfect red wax stamp pressed by an unknown hand centuries ago.

The air seemed to grow heavier still, vibrating softly with the weight of unseen watchers.

Mysty, perched lightly on a nearby crate, flicked her tail once.

"Well," she murmured, her yellow eyes gleaming, "it looks like the old man left more behind than just legends."

Kerilee cradled the fragile letter and scroll

against her chest, heart hammering.

Rip's final secrets had waited long enough.

Kerilee sat cross-legged near the center of the basement floor, the oil lanterns throwing long shadows across the stone walls. The others formed a loose circle around her, faces solemn, waiting.

With great care, she broke the brittle red wax seal on the envelope. The parchment inside was thin, almost translucent, but miraculously intact.

Her voice was steady as she began to read:

*"To my descendants,*

*If this letter finds thee, know that our line carries both a blessing and a curse.*

*Beware the demon Mahtantu, who feeds not on blood, but on the marrow of dreams. He is bound to the oldest corners of the earth, but his hatred of our kind festers still.*

*There exists a brew, taught to me by the leader of the Little Men, a draught brewed from Sassafras, night berries, certain roots, and sacred waters. It is no cure, but a shield.*

*I understood little of the full workings of his evil, but knew this: the curse was set not only upon my head, but upon all who would follow my name.*

*My hope was that time would erode his fury. I see now that it shall not.*

*Be wiser than I, braver than I, and seek the Breath of the Sky if all else fails."*

Kerilee's voice faltered slightly at the end, the words lodging in her throat.

The teens listened with rapt attention, the firelight dancing in their wide eyes. Jonameed clenched his hands into fists on his knees. Ruth sat perfectly still, as if memorizing every word.

Adrienna brushed the back of her hand across her cheek, blinking fast.

"He wrote like he feared no one would ever read this," she whispered.

Kerilee folded the letter back carefully, her hands trembling slightly now.

It wasn't just a relic.

It was a plea.

A promise reaching through the centuries, desperate to be fulfilled.

Mysty jumped down from her crate, landing silently beside Kerilee.

"About time someone started listening," she said softly.

For a moment, the weight of history, sorrow, and stubborn hope hung between them, thick as the damp autumn air.

And Kerilee knew—they were standing at the threshold of something much bigger than any of them had dared to imagine.

Kerilee set Rip's letter carefully beside her, then turned her attention to the scroll still bundled in waxed cloth. The fabric was cracked and brittle, barely holding together as she unwound it.

Inside was a parchment so aged it was almost golden, the ink faded but still legible under the lantern light. Symbols spiraled along the edges—stars, roots, drops of water.

Ruth knelt beside Kerilee, eyes shining with reverence. She traced the edge of the parchment without touching it.

"This... this is sacred craftsmanship. Not just words. It's part prayer, part protection."

Kerilee leaned closer and began to carefully read sections aloud.

The scroll described a recipe—a true brew of defense, gifted by the Little Men to Rip Van Winkle to shield his bloodline from the demon's touch. It spoke of brewing Sassafras bark with night berries, and adding very specific types of roots, singing certain names into the boiling mixture, and sealing the draught inside vessels blessed at sacred rivers.

Ruth's voice broke the silence:

"Have we found the full recipe before?"

Kerilee nodded slowly. "We did. It was with Tip and Sylvester's family documents. I've shown it to someone already—Arnoldus de Groot."

Adrienna's head snapped up, surprise flickering across her face. But after a moment, she nodded, trusting Kerilee's judgment.

"If anyone can help make it work... it's him."

Marcus, who had been quietly studying the top corner of the parchment, suddenly frowned. He pointed to a jagged symbol drawn above the recipe.

"Look," he said. "It's a warding mark. I know it from my uncle's lessons. It's meant to repel... something bad."

Jonameed leaned closer, his voice a low rumble.

"That symbol's tied to Mahtantu's true name. A name the old people were afraid to even whisper. They believed speaking it aloud would cause disease."

The basement seemed to grow colder at his words. A draft brushed Kerilee's skin like a ghost's touch.

Mysty's ears flattened slightly.

"Names have power," she said. "And some things don't like being reminded they were once chained."

Kerilee folded the parchment again, hands steady despite the chill that had settled in her bones.

The pieces were finally fitting together.

But the dangers were closer than ever.

Kerilee was about to set the scroll aside when her gaze caught on a section at the very bottom, almost an afterthought tucked beneath the recipe's intricate weaving of stars and vines.

She leaned closer, squinting at the faint lines of ancient script.

A poem—or maybe a riddle—curled there like the final breath of a dream:

*"She who danced on the Milky Path,*

*whose tears birthed the mountain springs,*

*holds the breath to banish fire."*

Kerilee's heart quickened. She recognized those words.

John Fisher's book on Sky Woman had spoken of a spirit whose dance spun the stars and whose sorrow shaped rivers and falls.

"She's real," Kerilee breathed. "And she's connected to Mahtantu's downfall."

Ruth, reading over her shoulder, touched the parchment with trembling fingers.

"That means... she's the key."

"She can remove him," Adrienna said, her voice barely a whisper, as if afraid to hope too loudly. "She can make him leave this world?"

Kerilee nodded slowly.

"But only if we summon her properly—and her blood descendants must call her."

Jonameed frowned. "Who are they, though?"

Everyone turned to look at Sheriff Embrey.

The Sheriff shifted, crossing his arms tightly across his chest.

For a long moment, he said nothing, only stared into the flickering lamplight.

Finally, he spoke, voice low and steady.

"My family. My people. My nieces and nephews."

Lucy, still clutching her silver whistle, blinked wide-eyed.

"You mean us. I knew it! We're finally ready. Thank you, Uncle Lenny."

Ruth's expression hardened into resolve.

"Then we'll do it. We'll call her."

Kerilee folded the scroll gently, a new weight settling on her shoulders.

The path ahead was dangerous—twisting through dreams, legends, and forces older than memory.

But for the first time, they had a true chance.

A gust of cold air stirred the lantern flames, making the shadows dance across the mural.

The world was shifting again—and there was no turning back.

A sudden buzz shattered the heavy stillness.

Sheriff Embrey pulled his phone from his belt, glancing at the screen. His face changed instantly—tightening, sharpening.

"It's Karyn," he said, voice clipped.

He answered with a brisk, "Embrey."

On the other end, Karyn's voice burst through, breathless and ragged.

"You need to come. It's Roger—he… he fell asleep again. We can't wake him—his lips are moving like he's speaking in some other language!"

A chill gripped the basement like an invisible fist.

Adrienna's hands flew to her mouth. Ruth stepped instinctively closer to the Sheriff, like she could anchor him through sheer will.

Kerilee's blood drained from her face.

"No," Mysty hissed under her breath, her fur bristling along her spine. "Not asleep. Not normal sleep."

"What do you mean?" Adrienna choked out.

Mysty's yellow eyes gleamed with grim certainty.

"He's been taken. Yanked into a time and place so deep... we might never reach him again."

Kerilee didn't hesitate.

"Get the whistle. Get the mural notes. We're going now."

Sheriff Embrey pocketed his phone and nodded grimly.

"Everyone outside. Fast."

The teens scrambled to gather their packs.

The old scroll and Rip's letter were bundled tight under Kerilee's arm as they raced up the narrow stairs, the trapdoor slamming open with a jolt.

Behind them, the mural's painted vines seemed to ripple in the lantern light—reaching, almost pleading.

And somewhere far beyond Garden Rock, an ancient blue fire roared in triumph.

# CHAPTER 15: SEND KERILEE TO SAVE ROGER

The rain whispered against the windows of Molenhaven General, the steady tapping a quiet counterpoint to the muted beeping of machines.

Kerilee stood just inside the doorway of Roger Van Winkle's hospital room, the fluorescent light making everything feel too stark, too sterile. The air smelled faintly of antiseptic and something sour underneath—the scent of fear trying to be scrubbed away.

Roger lay motionless in the bed, his chest rising and falling in a shallow rhythm. Wires and monitors clung to him like vines, recording numbers that meant nothing to the heart that watched.

Adrienna sat at his bedside, her fingers curled tightly around his hand.

"He won't wake up," she whispered, her voice raw with helplessness. "It's like he's trapped somewhere far away."

Mysty leapt lightly onto the bed, ignoring the rules of human spaces. Her yellow eyes narrowed as she studied Roger's face.

"This isn't normal sleep," she said, her voice low and vibrating with unease. "He's adrift."

Kerilee moved closer, feeling the wrongness in the air—like a book missing critical pages, its story left unresolved.

A nurse bustled past the open door, spotted Mysty perched on the bed, and stopped short with a gasp.

"You can't have animals in the hospital!" the nurse exclaimed, hands flapping. "You need to get it out of here immediately!"

Mysty sat up straighter, tail curling primly around her paws.

"But I'm a special service animal," she said sweetly, voice carrying just enough into the hall.

The nurse's mouth dropped open, a strangled noise escaping her throat. She blinked rapidly, glancing back and forth between Mysty and the humans as if trying to decide if she'd hallucinated the talking cat.

Adrienna sniffled, a tiny, broken laugh escaping her despite the tension.

Kerilee gently placed a hand on the nurse's arm. "It's alright. Mysty's... trained for emotional support. She doesn't leave his side."

The nurse nodded slowly, still staring at Mysty like she might sprout wings and fly away.

Outside the window, the rain thickened into a steady sheet, muting the world into shades of gray.

Inside the room, the only thing that mattered was the boy adrift between lifetimes—and the thin, unraveling thread that still connected him to this one.

The room's fragile calm broke when two more people arrived.

John Fisher entered first, moving with his

usual quiet gravity, as if he belonged to a rhythm slightly out of sync with the rest of the world. His sharp gray eyes swept the room, taking in Roger's still form, Adrienna's desperate grip on his hand, and Kerilee standing watch.

Behind him shuffled a woman bundled in a long brown cloak, her hair silver and plaited into thick braids that hung over her shoulders. Her face was lined deeply, but her eyes—there was something unmistakable about them. A brightness that time had not dulled.

The nurse, still lingering by the doorway and suspiciously eying Mysty, opened her mouth to object again. But John Fisher barely slowed.

He brushed two fingers gently across the nurse's forehead and murmured under his breath,

"I hear there's a new ice cream flavor in the cafeteria... glitterberry fudge ripple... only a few batches left..."

The nurse blinked. Her face transformed from confusion to sudden, singular obsession.

"I—uh—I have to go!" she blurted and dashed down the hall, shoes squeaking across the polished floor.

Mysty snickered. "Smart man."

John smiled faintly and turned to the rest of them.

"She'll be chasing that flavor for a good while."

Then the older woman stepped forward, her hands trembling slightly as she approached Roger's bedside.

She looked at him for a long moment—soft, aching.

"You don't remember me," she said, her voice low and thick with emotion. "But I've waited years to see you again."

Adrienna stared at her, confused.

The woman made her way carefully beside the bed and placed a hand over Roger's still hand.

"My name is Joanna."

Silence crashed through the room.

Kerilee's mouth parted, words failing her. Adrienna stiffened like she'd been struck.

Even Mysty's tail gave a sharp, startled twitch.

Older Joanna closed her eyes briefly, as if gathering herself against a tide of memories too deep to speak aloud.

"I promised I would find you again," she whispered. "And I have."

For a long moment, no one spoke. Only the low hum of Roger's monitors filled the room, an eerie steady rhythm counting seconds they could never recover.

John Fisher stepped forward, his voice steady, grounding them all.

"Mahtantu has thrown Roger farther back than ever before. He's not just asleep—he's adrift, stretched across time. Like a rope pulled so taut it's fraying."

Kerilee tightened her grip on her satchel instinctively, feeling the weight of Rip's scroll still inside.

Joanna, still beside Roger, lifted her chin.

"I tried to follow him back," she said, her voice trembling but determined. "We traveled together with others... for a time. We were in a desert land when it happened. Something went wrong. Roger felt responsible for me because of his dream curse. At first I enjoyed the attention, but as time wore on—I grew out of that. I discovered other things and other people. That altered things and Roger pulled away from me as I did from him. Then he vanished before out eyes."

Her hand shook as she brushed a lock of hair from Roger's forehead.

"I made it back through time. He didn't. I had

hoped he had returned here to his own time," Joanna murmured. "But, now I see that he is still lost out there."

Adrienna swallowed hard, her knuckles white around Roger's other hand.

"You mean he's... still traveling through time?"

"Worse," Mysty muttered from her perch on the foot of the bed. Her yellow eyes glowed solemnly.

"His cord to us must have snapped. That's why none of us could pull him back."

John's gaze sharpened on Kerilee.

"That's why you have to go after him."

Kerilee blinked, taken aback. "Me? Why me?"

John's mouth twitched, almost in a smile. "Because he'll know you, even when he doesn't recognize anyone else. And because you have something the rest of us don't."

"What's that?" Mysty asked, her tail twitching.

John's smile deepened, almost cryptic.

"Perspective."

Kerilee's pulse hammered in her ears.

"But are you coming too?" she asked him, almost desperately. "You know more about this than any of us."

John hesitated, then shook his head.

"I'll be there—but not in the way you expect. I exist... differently across time. In every place you'll visit, you'll find me. But you'll have to explain who you are each time. And why you need help."

*Kerilee's stomach sank. Alone in a sea of strangers. Each version of John a stranger too. How could she possibly explain everything in time, over and over? She swallowed hard, steadying herself.*

Kerilee stared at John, the reality of it sinking in like cold water.

She would be chasing Roger through splintered history—alone, except for those she chose to bring with her.

And John Fisher, for all his wisdom and mystery, could only guide from the margins.

Kerilee's mind raced, a dozen questions stacking behind her lips, but John held up a hand, his expression grave.

"There's more you need to understand first," he said.

The older Joanna turned from Roger and folded her arms, glancing once more at Roger's sleeping form. "He's drifting back and forth through moments... pivotal moments," she said. "Moments where John could still anchor him briefly."

"But that ability ends," John added quietly. "At a certain point his hope could be crushed, I cannot reach him if that happens. My influence fades, and Roger's soul would fall beyond even my grasp."

A chill slid down Kerilee's spine. "Fall where?" she asked.

"To a time where only my Master could save him," John said, his voice barely above a whisper.

Kerilee stiffened. "Your... Master?"

John nodded solemnly.

"You will know more when the time comes. For now, trust that you must not let Roger slip that far."

Adrienna's fingers tightened around Roger's hand.

"There's no way we're letting that happen. Tell us how to bring him back?"

John turned back to Kerilee. His gaze, so calm and ancient, locked with hers.

"I chose you because you're a librarian."

Kerilee blinked, taken off guard by the simple statement.

"A librarian?"

"Yes," John said with a faint smile. "You

understand timelines. Myth buried inside fact. Patterns lost to those who look only at the surface. You'll be able to recognize where you are in history. The others—"

He glanced at Adrienna, Mysty, and even Joanna—"might not. Not quickly enough."

Mysty hopped down from the bed with a flick of her tail.

"It's true," she said, her tone mock-grave. "The rest of us sucked in history class."

Adrienna gave a huffed laugh through her tears, grateful for the brief levity.

Kerilee pressed her hand to her chest, feeling the weight of her calling settle into place.

She wasn't the strongest. She wasn't the bravest.

But she knew the importance of stories—and the danger of forgetting them.

Slowly, she nodded.

"If I go... how do I find him?"

John reached into his coat and pulled out a pendant—a simple thing at first glance, a teardrop-shaped stone set into a battered silver frame. The surface shimmered like a mirror caught between one breath and the next.

He held it out to Kerilee.

"This will follow his soul-print," he said. "It will resonate when you're close to him. Or he to you. Even across centuries. Your mind will subconsciously be scanning through where he might be in time. When you get a strong enough hit, you and the others will stop in proximity to where and when he is. It is complicated to explain, but the pendant will help."

Kerilee accepted it carefully, the pendant warm against her palm. It pulsed faintly, almost like a heartbeat not her own.

Mysty eyed the pendant critically. "Let's hope it doesn't lead us straight into a dragon's mouth."

Kerilee snorted. "Don't jinx it."

John stared at Mysty, "I'm sure Mysty is just trying to make light of the situation. You won't be lead into a creatures mouth. At worse, you may appear in a puddle or some distance apart, but you will eventually find each other and interact with both Roger and myself."

Adrienna rose from her seat, squaring her shoulders.

"Then I'm coming too," she said, her voice steady and unshakable.

Mysty flicked her tail sharply.

"As am I. No way you're wandering time alone.

You need someone quick-witted and charming."

"And sarcastic," Adrienna added dryly, managing a small smile.

John nodded approvingly.

"You'll need each other."

Older Joanna reached into the folds of her cloak and produced something that made everyone lean forward: an old iron nail, large and stained dark with something long since dried.

"This was in his hand the last time I saw him," Joanna said, her voice trembling. "It's old. Ancient, maybe. From somewhere... not here. And he was quite shaken up, heartbroken. That was the moment he vanished."

Kerilee took it, feeling the heavy weight of history cling to it.

Mysty sniffed the nail suspiciously.

"Not cursed. But definitely connected."

John watched them all quietly, the shadows deepening around his features.

Joanna hesitated, then said, "John said the plan was to have you find Roger in time before that happens. Then give him a reason to return with you."

Adrienna spoke up. "Oh! I'll give him reasons to come back. Don't you worry about that!"

Kerilee tightened her grip on the pendant, heart pounding.

"We'll bring him back," she promised.

John stepped closer, his voice softer now, almost personal.

"You may learn certain truths about me in your travels," he said. "Things not meant for this time. If you happen to piece together who I really am... please. Keep it to yourself."

Kerilee met his gaze evenly, feeling the unspoken weight of the promise.

"I will."

John smiled faintly, a sadness flickering behind it.

"Then it's time."

The rain pounded harder against the windows as if the storm outside recognized the storm in time about to begin.

And somewhere, across tangled threads of history, Roger Van Winkle waited—lost, adrift, and running out of time.

# CHAPTER 16: IT'S TIME FOR TIME

The rain pattered steadily against the Diner's back window, threading silver rivers down the glass. Inside, the room was dim, the overhead lights off, the only glow coming from a small lamp in the corner that buzzed faintly. It had already been a long day with the Sheriff's family, and meeting with John and Joanna at the hospital. John had already given further instructions for our sort of time travel. Apparently, there are many ways to approach what you target in time. Sometimes people target a year, or an event. We were targeting a person which is why the pendant was needed. John had to sort of program the pendant to each of us by having us touch it, and then touching the pendant to John's hand to target him. Once that was done, those of us that were going met at the diner. Older Joanna's touch was entered into the pendant differently since it was her younger self we were to find. Older Joanna would remain in John's care until we got back.

Mysty yawned. "Sorry, Ladies, but this storm must have some mega-electrical interference that's messing with or delaying our trip back through time."

Kerilee sat cross-legged on the floor, the cold from the linoleum seeping through her jeans. She turned the mirrored pendant slowly between her fingers, watching the light bend and ripple across its surface. It didn't pulse yet. Not here. But soon.

Adrienna sat nearby, elbows braced on her knees, clutching the bloodied iron nail as if it were an anchor. Her knuckles were white, and she stared past it—past the room entirely, it seemed—toward someplace only her heart could see.

Mysty perched on the windowsill, her silhouette outlined against the wet, flickering world outside. A flash of lightning illuminated her for a moment, her yellow eyes gleaming.

The silence stretched, taut and brittle.

Kerilee broke it first. "Are you sure you're okay to do this?" she asked Adrienna softly.

Adrienna didn't look away from the nail. "We're going to save him," she said, her voice raw but firm. "That's all that matters."

Kerilee nodded, though worry gnawed at her chest.

Mysty stretched luxuriously, paws flexing against the wooden sill. "We're chasing a boy through centuries," she said. "This is either courage or chaos."

Kerilee gave a humorless chuckle. "Maybe a little of both."

Adrienna suddenly grimaced and rubbed her temples. "Ugh. My head feels like it's been wrung out."

She pushed herself up unsteadily and called, "Karyn?"

A moment later, Karyn Molen bustled in, still wearing her apron and wielding a spatula like a scepter. "You girls need something?"

Adrienna managed a weak smile. "Sorry, Karyn. I hate to bother you, but... headache."

Karyn's eyes softened immediately. She turned to a cluttered shelf near the sink, rummaged around, and tossed a small brown bottle toward Adrienna with practiced ease.

"Catch."

Adrienna caught it with one hand, blinking at the label. "Willow bark extract?"

"Old-fashioned, but it works, "Karyn said. Her voice lowered, heavy with meaning. "Bring my Roger—my almost nephew—back safely."

Adrienna clutched the bottle to her chest, nodding wordlessly.

Karyn gave them a tight, supportive smile and disappeared again, the clatter of dishes fading behind her.

The rain drummed harder against the window.

Kerilee looked at the pendant once more. The glass shimmered—briefly—as if it too were waiting.

They settled onto sleeping mats Karyn had

hastily tossed into the room—thin, worn things that smelled faintly of old flour and sassafras wood.

Kerilee lay back, staring at the cracked ceiling tiles, the pendant clutched loosely in her hand.

After a long silence, she couldn't help herself. "Who do you think John Fisher really is?" she asked, her voice barely more than a whisper.

Adrienna sighed heavily, rolling onto her side. "Focus, Kerilee. Roger is who we should worry about right now."

"But think about it," Kerilee pressed, ignoring the warning tone. "Someone that powerful... that old... walking around in our time? Doesn't that make you curious?"

Mysty yawned extravagantly, flashing sharp little teeth. "Too many Johns in history to figure it out now," she said. "John Smith, John the Apostle, Johnny Appleseed... Take your pick."

Kerilee huffed, unsatisfied. She turned the pendant over in her fingers again, feeling its quiet, waiting hum against her skin.

"But how many Johns never died?"

Mysty sat up, her yellow eyes flashing in the low light. "Let. It. Go," she hissed. "If he's who I think he is, then he's got a good reason—

and I mean an extremely good reason—to stay anonymous."

Kerilee flinched slightly under Mysty's fierce stare. "Okay," she relented. "I'll drop it. For now."

"Thank goodness," Mysty said, flopping back onto her mat with theatrical exhaustion.

The rain's steady drumming deepened, each gust of wind rattling the old window frames.

Kerilee closed her eyes, but her mind refused to settle.

The mysteries surrounding John Fisher would have to wait. First—they had to find Roger.

Kerilee shifted restlessly on her mat, sensing the weight of what was coming—what was already pulling them.

When she sat up, she noticed something odd. A fine dust coated the floor, forming a wide, imperfect circle around their sleeping mats. She brushed her fingers through it and realized it wasn't dust at all.

It was flour.

Mysty padded over, paws sinking lightly into the powder. "Ooh," she purred, flexing her toes. "I love how this feels on my paws."

Adrienna sat up too, squinting at the floor. "What's all this?"

Before Kerilee could answer, Karyn poked her head through the door again, wiping her hands on a dish towel. "I didn't have enough salt," she said sheepishly. "So I used flour."

Kerilee gave her a bewildered look. "Karyn, you've got enough Sassafras wood in this back room to keep seventy demons away from this place!"

Karyn shrugged with a grin. "Figured it couldn't hurt to have a little extra."

Kerilee sighed. "It wasn't necessary—but knowing Karyn, she wasn't taking chances."

Adrienna laughed under her breath and shook her head, settling back down on her mat.

Mysty twirled in a slow, delighted circle, leaving crisp little pawprints in the flour. "Not complaining," she said. "Feels like a fancy spa treatment."

Kerilee smiled faintly but sobered quickly when Mysty's ears twitched back, alert.

"We'll miss the first few crucial stops because of the delay, "Mysty warned, voice low. "Brace yourselves—we're starting in turbulence. The storm's interference—makes us skip ahead. We won't know what Roger endured at the start."

Kerilee's fingers tightened around the pendant

instinctively.

Outside, the storm deepened, the windows rattling harder now, as if some unseen hand beat against them, demanding to be let in.

Inside the circle of flour, three lives braced to plunge into history's current—ready or not.

The three of them lay down, their heads pointing toward the center of the flour circle. It felt oddly ceremonial, like some long-forgotten rite reawakening in the flickering shadows of the back room.

Kerilee adjusted the pendant against her chest, feeling its mirrored surface pulse faintly, like it could already sense the thinning boundaries between worlds.

Rain hammered the roof harder now, the sound thick and insistent, blending with distant thunder that rumbled low and long—like a drumbeat from the belly of the earth.

Kerilee whispered into the charged darkness, "Hold on, Roger. We're coming."

Adrienna, lying stiffly beside her, gripped the bloodied nail tighter and muttered through clenched teeth, "I'm not losing him again."

Her voice cracked, and Kerilee heard the truth hiding under the bravado: Adrienna was terrified.

Mysty's tail flicked once in the flour as she stretched luxuriously. "And if we find Mahtantu's footprints along the way," she said, her voice a velvet growl, "I say we scratch his eyes out."

Kerilee smiled tightly, grateful for the defiant humor even now.

A deeper rumble of thunder shook the diner's walls, and the windows vibrated in their frames.

Mysty cracked one yellow eye open. "One last thing..."

Kerilee and Adrienna both turned their heads slightly toward her, listening.

"The circle's delay... we'll miss the first stops. Whatever Roger faced—we won't see it."

Adrienna tensed visibly, but said nothing.

Kerilee's stomach twisted with new dread. They would be chasing echoes—shadows of events they hadn't witnessed but would still have to undo.

The pendant in Kerilee's hand sputtered with a small flash at first. They each looked at each other briefly when the pendant flared suddenly—bright and cold like a miniature lightning bolt.

Then silence.

Their bodies slackened on the mats as their

spirits slipped free—drawn into the vast current of time, falling away from the rain, the diner, and everything they knew.

The flour circle remained undisturbed, save for the faint imprint of a black paw pressed into the very center.

# CHAPTER 17: DIMWITTY DANGER

IF WE IS ANY STUPIDTER, WE'D BE REALLY DUMB.

Fog coiled over the fields like sluggish ghosts as Sheriff Len Embrey pulled his unmarked SUV onto the Dimwitty farm's gravel drive. The tires crunched over broken stones and dead leaves, headlights pushing back the gray only a few feet ahead.

He shifted the vehicle into park but left the engine humming low. In the seat beside him, Deputy Marla Sanders checked her sidearm, while in the backseat, Ruth sat rigidly, her dark braid tucked beneath the hood of her jacket.

Len flipped through the thin case file resting on the dash—a handful of photographs showing strange deliveries made to the Dimwitty Farm at odd hours, glowing lights spotted through the woods, and a growing list of whispered rumors: relic-trading, spell-selling, secret rites tied to the old blue demon. Mahtantu's fingerprints were all over it, if you knew what to look for.

He glanced at Ruth. "Stay close. Keep sharp."

She nodded, her breath fogging the window as she leaned closer, scanning the farmhouse through the swirling mist.

"Something under that house stinks of old magic," she muttered.

Len's jaw tightened. He didn't like bringing someone so young into something this deep, but Ruth wasn't just another teenager. She was trained for this—prepared in ways few others could be.

He killed the engine and opened the door. Cold damp air rushed in, curling against his skin like a living thing.

Time to poke the bear. Or in this case, the morons pretending to be human.

Len led the way up the sagging porch steps, his boots landing heavily on warped planks that moaned under his weight. Deputy Sanders flanked him. Ruth kept a few careful paces behind, her eyes scanning everything—the windows, the darkened barn nearby, the shadowed tree line beyond.

The front door swung open before Len could knock.

Lester Dimwitty stood there in a grease-stained flannel shirt and ripped jeans, his grin too wide, too eager.

"Well, well," Lester drawled. "If it ain't Sheriff High-and-Mighty himself. To what do we owe the pleasure?"

Before Len could answer, another figure shuffled into view—Cobb Dimwitty, broader and thicker than his brother, his hair a tangled mess, his smile slick and slow.

"Come on in, Sheriff," Cobb said, stepping aside. "Always got time for visitors." He held up a chipped mason jar filled with something that sloshed dangerously between clear and amber. "Drink?"

Len kept his face neutral. He stepped inside

without touching the offered jar.

"Routine check," he said casually. "Some folks been talking about strange happenings out this way. Thought we'd swing by."

Cobb chuckled, low and wet.

"Strange? Round here? Naw. Just honest farm work." His eyes glittered with something colder than humor. "You city types get spooked easy."

Lester leaned against a battered side table, arms crossed.

"Or maybe," he said, voice thick with mockery, "you're hearin' about the blue demon. Scary bedtime stories for old men."

Len's stomach tightened, but he only raised an eyebrow.

"You think that thing works for you?"

Both brothers barked out harsh laughter. Cobb set his jar down with a clink.

"Work for us?" Lester hooted. "Sheriff, you've got it backward."

He leaned in, breath sour with drink and something fouler.

"We work for him."

The room seemed to tilt slightly, as if the walls themselves recoiled from the truth.

Ruth, standing quietly by the door, lifted her

phone slightly and tapped the record button.

Len kept his voice even, but the steel behind it sharpened.

"Care to explain that?"

Cobb's smirk widened. "Wouldn't matter if we did. You're already part of it, whether you know it or not."

Len's hand drifted casually toward his belt, where his weapon hung.

Just in case.

Because whatever game the Dimwittys were playing—it wasn't just rumors anymore.

Len let the brothers' laughter echo behind him as he caught Ruth's subtle nod toward the back of the house. Cobb and Lester were too busy congratulating themselves to notice when Len shifted his weight and jerked his chin toward Deputy Sanders.

Without a word, they moved.

Down a narrow hallway lined with peeling wallpaper and water-stained photographs, they found a heavy rug thrown carelessly across the warped floorboards.

Len tapped the rug with his boot. "These old houses," he muttered. "Always hiding something under the floorboards.'

Ruth crouched and tugged at a corner. Dust plumed into the air, making her cough. Beneath it, plain and ugly, was a rusted iron ring set into a narrow trapdoor.

Len knelt beside her, gripped the iron ring, and heaved. The hatch groaned like an animal in pain as it lifted. From the pit below, a wave of sour, throat-burning stench surged up—a smell like old death, curling around them as if something below had just taken a breath.

"That's not natural," Ruth whispered, covering her nose with her sleeve.

Len swung the trapdoor open fully. A rough, earthen staircase spiraled down into darkness, lit only by the faint blue flicker of something burning below.

He exchanged a look with Deputy Sanders. She nodded once, slipped her weapon free, and swept her flashlight ahead, the beam trembling faintly as she moved into the dark.

One by one, they descended.

"Oh, Sheriff," Lester called from above. "We really hope you find what you're looking for."

The basement sprawled larger than it had any right to be, hollowed by generations of digging hands. Rough stone walls jutted with strange relics—charred bones wedged into cracks, ancient coins pressed into mortar, rusted tools dangling from crooked nails, each one pulsing faintly in the unnatural blue light.

At the chamber's center, a brazier burned with unnatural flames—the color of glacier ice—and around it stood a circle of crude pedestals. On one lay an open parchment, blackened with age, marked by a sigil that twisted if you stared too long. Mahtantu's mark.

A golden chest rested nearby, its lid thrown open. Inside glittered offerings: thick gold coins bearing tribal carvings, bloodstained rings, and weathered artifacts older than memory.

Ruth edged closer, her voice a harsh whisper.

"They've been paid in blood and gold... to keep chaos alive."

"Len swallowed hard, his eyes tracking every jagged bone and glint of gold like stepping stones into a grave."

This wasn't just superstition.

This was loyalty bought and paid for—and Mahtantu wasn't done collecting.

At the far end of the chamber, something caught Len's eye—a stretch of wall, rougher than the others, smeared with soot and faded paints.

He moved toward it, raising his flashlight.

Len's flashlight peeled back the mural's shadows, line by line: a family tree, sprawling and grotesque. Each generation of Dimwittys was depicted as stick figures with burning hands, flames licking up their arms and shoulders. Names were scrawled beneath them —most nearly illegible—but the line traced all the way down to Lester and Cobb, their faces caricatured in cruel grins.

At the very bottom, pinned awkwardly beneath the final branch, was something else.

A photograph.

Sheriff Len Embrey stared at it in stunned silence. It was an old department photo— himself, fresh-faced and proud, standing beside his first patrol car. Someone had scrawled across the bottom in thick, red ink: INTERFERER.

Ruth sucked in a breath beside him.

"They've been keeping records on you. And me," she said, pointing to the margins where lists of names ran down the wall—some crossed out, some underlined, some marked with strange sigils.

Len's fists curled tighter at his sides.

He pulled out his phone and began snapping

pictures—of the mural, the burning family tree, the photograph with its damning word.

"This," he said, his voice low and tight, "needs to be taken down."

Deputy Sanders joined them, her face pale as she took in the room.

"This isn't just loyalty. It's a legacy," she said grimly. "Born, bred, and passed down."

Len nodded. He could feel it now—heavy as the damp air around them.

This wasn't just about fighting a few reckless Dimwittys anymore.

It was about uprooting an evil that had been sown into the very blood of Molenhaven.

They closed the trapdoor and moved fast, hearts hammering.

But when they returned to the living room, Lester and Cobb were gone.

Deputy Sanders cursed under her breath and radioed for backup.

Len scanned the room—no sign of a fight, no overturned chairs, no broken glass. Just... absence. Too neat. Too sudden.

"Fan out," Len ordered. "Find them."

Ruth was already moving, slipping through the mist-choked backyard. She followed a faint, broken whimper—a sound too soft, too human to ignore—leading her toward a dilapidated shed sagging at the edge of the property.

She hesitated only a moment before reaching for the door. A prickle ran up her spine as she neared the door—like she was stepping over a threshold she wasn't meant to cross.

Len opened his mouth to call her back—but the mist swallowed her before the words formed.

It creaked open on rusty hinges.

A scream shattered the mist-heavy silence.

Len sprinted outside, weapon drawn. Sanders was right behind him, both of them racing toward the sound.

But by the time they reached the shed—

Ruth was gone.

Her cracked phone lay in the dirt, its screen flickering with static. No footprints. No scuffle. Just the mist swallowing the ground, and a silence that rang in Len's ears.

He bent and picked up the phone. As he turned it over, the screen jolted to life one last time.

Ruth's voice—breathless, terrified—burst through the speaker:

"He's erasing my mind—Sky Woman—STOP HIM!"

A man's voice shouted in the background: "Cobb, stop her!"

A muffled scuffle, the sound of something hitting the floor, a sharp gasp—

Then nothing.

Static hissed from the phone's speaker, filling the silence like ash.

Len stood frozen, the phone heavy in his hand, the words searing into him like hot iron. The fog pressed closer, thick and smothering. He felt it in his bones now—the weight of what they'd just lost.

Grief clawed up his throat, but he swallowed it down, forcing his breathing steady.

He turned slowly to the others, his voice rough with fury.

"He's just made it personal."

# CHAPTER 18: WHO SUMMONS SKY WOMAN NOW?

Sylvester Van Winkle wiped the blade of his whittling knife against his jeans and inspected the sassafras branch in his hand. Smooth now. Ready for whatever project he felt drawn to next. Something was about to happen. He could feel it, same as a man could feel a thunderstorm in his bones.

The heavy wooden doors of the barn creaked

open, letting in a gust of cold air—and Sheriff Embrey.

Sylvester straightened up. The Sheriff wasn't wearing his hat, and that alone made Sylvester's gut tighten. The man always wore his hat.

"She's gone," Sheriff Embrey said, voice raw. "Ruth's gone."

The sassafras branch slipped from Sylvester's fingers, landing with a dull thud on the dirt floor. "What do you mean gone?"

Before the Sheriff could answer, a group of teens—wild-eyed and panting—rushed in behind him. Ruth's cousins and brother: Jonameed, Phil, Lucy, and Marcus. A scrappy, stubborn crew that normally raised more dust than a cattle stampede. But tonight, their faces were pale. Set.

"They took her," said Jonameed, fists clenched at his sides. "We think it was the Dimwittys."

Sylvester narrowed his eyes. "Dimwittys?" His voice came out low, dangerous. "You're sure?"

"They were lurking around town earlier," Jonameed said. "Acting jumpy. Then Uncle Len took her and a deputy to talk to the Dimwittys. Ruth went missing while they were there. We found this by the road."

He held out a scrap of Ruth's shawl which she wore under her jacket, caught on a broken branch, stained with something dark that Sylvester didn't want to examine too closely.

Sheriff Embrey scrubbed his face, aging in front of them. "We got tracks leadin' toward the old woods. Too many to follow clean. We need help."

Sylvester's jaw tightened until his teeth ached. The Dimwittys had always been trouble, but if they'd taken Ruth, it wasn't mischief anymore. It was war.

He bent down, picked up the sassafras branch from the floor, and tucked it into his belt. His voice was steady when he spoke.

"Then let's not waste any more time."

The moon was a thin blade in the sky as they set out, its cold light slanting through the naked trees. Grabbing his walking stick, Sylvester led the way, his boots crunching over fallen leaves, his steps sure despite the uneven ground.

The Sheriff and the teens—Jonameed, Phil, Lucy, and Marcus—followed close behind, their breath misting in the chilly air. No one spoke. The forest pressed in around them, thick with old secrets.

Half an hour in, the trail thinned to a ghost of itself, little more than a game path. Sylvester paused, cocked his head. A low sound drifted through the trees—the mournful hoot of an owl.

"Hear that?" he murmured.

Another hoot. This time, eerily shaped into something more: Ru-u-uuth.

The Sheriff swore under his breath. Marcus stumbled and grabbed Lucy's arm, his eyes wide.

"Keep moving," Sylvester said. "I guess the owls loved her too."

They pressed on, the air growing thicker, the shadows deeper. Sap oozed from ancient trees, glowing softly in the moonlight.

"Never seen that before," Phil whispered.

"Neither have I," Sylvester admitted. He touched a glowing drip—it was warm, and it thrummed faintly under his skin.

Finally, the path ended at a clearing ringed in thick, velvet moss. A flat stone altar stood at the center, worn smooth by time.

Sylvester felt the hairs on the back of his neck stand up. This was it. The Secret Place.

He turned to face the Sheriff. "I thought we were following the Dimwitty tracks?"

Len opened his mouth to speak, then he let out

a shaky breath. Phil answered for him, "We were following them, but they led here."

"Why here?" said Len. "We can't summon Sky Woman without Ruth!"

Without needing to be told, Jonameed, Phil, Lucy, and Marcus knelt around the stone, heads bowed.

Sylvester stepped forward. He placed his broad palm flat on the altar's cool surface. The stone pulsed once—like a heartbeat—and the air grew heavy, charged with unseen energy.

He closed his eyes and whispered, "We come seekin' help. We come seekin' hope."

The ground beneath them shivered. A light began to descend.

The light sharpened as it touched down, weaving itself into the shape of a woman—tall, ageless, draped in a robe of mist and moonbeams. Her hair spilled loose down her back, and woven among the dark strands were tiny, glimmering stars that twinkled as if alive. Len looked with incredulity at Sylvester. How was he able to do this?

Sylvester swallowed hard. He wasn't a man to spook easy, but something about her presence made his heart pound in his chest like a wild drum. Sacred. Powerful. Kind.

The woman's eyes, deep as midnight oceans, settled on him. When she spoke, her voice was both thunder and lullaby.

"You carry the pain of many," she said, her gaze piercing straight through him. "And yet, you still come with hope."

Sylvester found himself rising to his feet without willing it. His voice was steady, but low, respectful. "Are you... Sky Woman?"

A small smile curved her lips. "Names are like river stones, Sylvester Van Winkle. They change with time and current. But yes—once, I walked among your kind."

The teens behind him gasped quietly. Sheriff Embrey dropped to one knee, bowing his head.

Sky Woman drifted closer, though her feet never seemed to touch the moss. The clearing pulsed softly with her every movement, like the world itself was breathing her in.

She turned her face toward the Sheriff, giving him a look of aching tenderness.

"I once walked with your people," she said. "Long before the names you now carry. Long before towns and roads carved the land. I was chosen to watch over them—from a place beyond stars."

Sylvester frowned slightly. "Sheriff's Mohican, right?"

Sky Woman's eyes twinkled with sadness and something deeper.

"Not the tribe he's currently in. I meant the tribe that lived here more than two thousand years ago. We are all threads from one of the twelve great tribes. You, me, him—our blood remembers even when our tongues forget."

Sylvester planted his feet, the burden of generations anchoring him in place.

Sky Woman's gaze grew darker. Heavy.

"As for the one you call Mahtantu," she continued, "he is not new to wickedness. Long before he tormented the Van Winkles, he tormented my people. He shattered timelines, corrupted hearts, and wove despair into the roots of the world."

The clearing seemed to dim at her words. Even the glowing sap dimmed to a flickering, uneasy light.

Sylvester felt his jaw tighten. "Can you stop him? Can you... take Mahtantu away?"

The others looked up, desperate hope burning in their faces. Even Sheriff Embrey, tough as ironwood, turned his gaze toward Sky Woman like a man reaching for the last rung of a crumbling ladder.

Sky Woman's expression softened, but there was sorrow in her eyes.

"I cannot," she said gently. "I am bound to watch, to guide, but not to interfere." She knelt down herself, robes pooling around her like liquid silver. "But I can call to those who once imprisoned him. They dwell beyond the rim of your world, and they are listening."

Sylvester shifted, the damp moss cool against his boots, grounding him like an anchor in a rising tide. "How long will it take?"

Sky Woman's voice was heavy with both wisdom and warning. "Their paths bend slowly through the stars. They will come—but not swiftly. Time here and time there do not move the same."

Sylvester set his jaw. Waiting wasn't in his nature. Especially not with Ruth's life hanging in the balance. "And what do we do until then?"

Sky Woman's eyes gleamed, catching every stray glint of starlight. From a hidden fold of her robe, she produced a tiny pouch woven from what looked like spider silk and morning dew.

She placed it in Sylvester's outstretched hand.

"Make ready. Heal what you can. Fight if you must."

Sylvester looked down at the pouch, skeptical. "Seeds?"

Sky Woman's smile deepened. "Drop one on the ground beside you. See for yourself."

The Sheriff and teens leaned closer, breaths held. Even the woods seemed to hush, the trees leaning inward.

Sylvester pulled the drawstring open with a rough thumb and forefinger. Inside, nestled in the shimmering fabric, were a dozen tiny seeds, each glowing faintly like captured fireflies.

He plucked one out and, after a quick glance at Sky Woman, dropped it beside his boot.

The moment it touched the earth, the ground trembled—and from the moss sprouted a tendril, unfurling like a time-lapse miracle. A plant rose in seconds, thick-stemmed and crowned with a single glowing fruit the size of his fist, its skin a blend of gold and crimson.

Sky Woman inclined her head. "Take. Eat."

Sylvester hesitated a beat—then shrugged. "In for a penny." He plucked the fruit free, its skin warm against his fingers, and took a bite.

The taste was unlike anything he'd ever known—sweet, rich, alive. The moment the juice hit his tongue, something ignited inside him. He staggered backward, clutching his leg out of reflex—except... the ache, the slow grinding pain he had lived with since the attack, was gone.

He bent his knee, testing it. No pain. No stiffness. His breath caught as strength flooded into places that had grown weak over years of quiet suffering—old scars fading, his joints moving freely, the dull roar of exhaustion lifting from his chest. He stood straighter. Taller. Younger.

Sheriff Embrey let out a low whistle.

Jonameed's voice shook. "It's like he's been... reborn."

Sky Woman watched silently, her eyes full of ancient knowing.

Before Sylvester could speak, the clearing shimmered—and a new vision tore across the sky like a ragged hole in reality.

The mossy glade seemed to freeze as the vision opened, hanging in the air like a window torn through time.

There was Ruth—her black hair matted and wild, her wrists bound tightly behind her back with rough rope. She sat on the dirt floor of a room in shambles, whispering to herself, rocking slightly.

Standing over her were Lester and Cobb Dimwitty, their faces twisted with glee and something darker. Lester held a glowing object in one hand, sharp-edged and pulsing with a sickly green light. Cobb clutched a heavy mallet, testing its weight with little swings that made Ruth flinch.

Sylvester stepped forward instinctively, but his hand met only air—the vision was untouchable.

Sky Woman's voice echoed through the clearing:

"Your time is short."

The vision trembled and faded.

Sylvester turned to the Sheriff and Ruth's cousins.

No words.

They ran.

# CHAPTER 19: THE FIRST STOP IN TIME

A sharp crack split the air above Kerilee's head. She jerked awake with a gasp, the tang of smoke and damp earth clogging her senses. Overhead, the towering limbs of a maple tree stretched against a bruised blue sky, the late afternoon light streaking the woods in patches of gold. Somewhere beyond the trees, musket fire rattled like hail on a tin roof.

Kerilee sat up slowly, her head swimming.

Adrienna groaned beside her, shielding her eyes from the sunlight, while Mysty let out a low, grumpy yowl and shook dirt from her fur.

The air itself seemed to hum, and Kerilee's skin prickled with static, as if she were shaking off water—or centuries.

"What happened...?" Adrienna muttered, brushing leaves from her jeans—except they weren't jeans anymore. Kerilee's heart stumbled. Her clothes had changed—gone were her sneakers and hoodie. Now she wore rough homespun cloth, her skirts dusted in mud. "Hey, where did the bloody giant nail go?"

"We're... not in Molenhaven anymore," Kerilee said, her voice barely a whisper. "Maybe something changed just by coming here. Just in case, let's keep our eyes open for it."

A stream gurgled somewhere nearby. Through the trees, movement caught her eye—a young man, his clothes straight out of a history textbook: breeches, a linen shirt, a battered tricorn hat tilted jauntily over his brown hair.

"Roger!" Kerilee cried, scrambling to her feet. He waved from across the stream, a relieved grin stretching his face.

"How did you girls get here?" he called, wading

ankle-deep through the water toward them.

"John gave us a tracker," Kerilee said. "Who changed our clothes?"

Roger pointed toward a nearby house. "The maids brought out these clothes when I couldn't wake you. They changed you both and gave me this stuff to wear. I told them what you had on was fancy underwear from Paris—they weren't impressed. Said they'd have your things cleaned by morning."

Kerilee pressed a hand to her chest. The world around her smelled of churned soil and woodsmoke, sharp and real. The air was cooler, crisper, and buzzing faintly with distant hoofbeats.

"This isn't just a dream," she said, breathless. "It's history."

Mysty leapt onto a fallen log, flicking her tail. "And as usual, we're fashionably late."

Adrienna stomped across the uneven ground, skirts swishing angrily around her legs. She planted herself in front of Roger, arms folded tight across her chest.

"Care to explain why you've been ghosting me through centuries?" she demanded, her voice low and sharp.

Roger opened his mouth, but before he could speak, a shout rang out. A girl came sprinting from the nearby house—a young Joanna, her hair pinned up under a lacy cap, her blue dress billowing as she ran. Without hesitation, she threw her arms around Roger and clung to him as if afraid he'd vanish again.

"I thought I lost you again!" Joanna gasped, her voice breaking.

Roger patted her back awkwardly, casting an apologetic glance toward Adrienna, whose jaw clenched hard enough to crack a stone.

Kerilee hurried forward, sensing the emotional powder keg about to ignite. She touched Joanna's arm gently to get her attention.

"Joanna, listen to me," Kerilee said urgently. "We're from your future. You're seventy now, and you sent us back here to help Roger."

Joanna leaned back, confusion creasing her forehead. "That... that makes no sense. I'm right here."

Mysty, perched on Adrienna's shoulder, snorted. "Time rarely does. Best to just roll with it, kid."

Joanna's eyes darted between them, searching for any sign of a prank. Her hands fidgeted with the apron tied around her waist.

Kerilee pressed on. "Roger, do you have any idea why we didn't meet you sooner in time? Where else have you been?"

Roger rubbed the back of his neck, avoiding Adrienna's furious stare. "First, I landed by a river in Pennsylvania. It was freezing cold—early spring, I think. After that, I woke up on a hill in western New York. No towns, no people—just grass, log cabins, and wind for miles."

"And now here," Adrienna said tightly, her arms still crossed.

Roger gave a lopsided shrug. "I guess the third time's the charm?"

A commotion near the tree line drew their attention. Soldiers in rough colonial uniforms moved in formation, the glint of bayonets flashing between the trees. Standing a little apart from them, speaking intently with an officer, was a man wearing a weathered tricorn hat and a rough spun vest.

Kerilee's breath caught in her throat. There he was—the man who haunted dreams and whispers, standing flesh and bone beneath colonial banners. "That's him," she whispered. "John Fisher."

Roger cupped his hands around his mouth and called out, "John! Over here!"

The man turned, wary. His sharp eyes swept across the group—lingering briefly on Kerilee, Adrienna, and Mysty with a frown of confusion. He approached slowly, every step measured.

"Who are you?" John asked, voice edged with suspicion.

Kerilee stepped forward, heart pounding. "We know you," she said. "In the future." She held up the pendant. "You gave me this to find Roger."

John stiffened. His hand went instinctively to the hilt of the knife tucked into his belt.

"Then this," he said carefully, "is the moment my memory splits. I have not yet lived that life."

Adrienna moved to stand beside Kerilee. "We're here because Roger got stuck," she explained. "Joanna's future self came to us for help."

John regarded them for a long moment, eyes hard as flint. Then, with a slow, reluctant nod, he said, "So this is where the threads begin to knot."

Kerilee met his gaze, steady and sure. "We are to follow your thread back in time, John."

For a brief second, something almost like sadness flickered in John's expression before he shuttered it behind a stoic mask. He inclined his head, but his posture stiffened. He didn't look pleased about the arrangement.

Mysty muttered under her breath, "Great. Another bundle of sunshine to add to the group."

As the soldiers dispersed back toward town, the group found a quiet patch of woods. Roger leaned against a crooked tree, running a hand through his messy hair. Adrienna lingered nearby, arms folded, her body tense with unsaid words.

Finally, she marched up to him, stopping just shy of arm's reach.

"You said she was just a dream," Adrienna said, her voice tight with hurt. "All those nights—you made it sound like she didn't matter."

Roger shifted uncomfortably. "She was," he said. "I mean—she wasn't real. Or I thought she wasn't. I didn't ask to see her every night. I didn't choose it."

Joanna, who had been lingering a few paces behind, heard him. Her face, open and hopeful a moment ago, crumpled. She flinched as if struck.

Joanna's voice turned bitter. "Once again—where does that leave me?"

A heavy silence fell. Even the birds seemed to hush.

Kerilee stepped forward, planting herself firmly between them before the tension could snap.

"This triangle is Mahtantu's doing," she said, her voice low but sure. "It's the byproduct of a curse. A curse that feeds on confusion, pain, and tearing people apart. People like you, Roger, and Adrienna."

She glanced toward Joanna too, making sure none of them felt spared from the truth.

Mysty, perched lazily on a fallen branch, flicked her tail. "And hormones," she muttered just loudly enough to earn a sharp glare from Adrienna—and a half-hearted snort from Roger.

Kerilee crossed her arms. "The only way we break this is if we stop letting the curse use our emotions against us."

For a long moment, nobody spoke. Roger stared down at his boots. Adrienna hugged herself tighter. Joanna blinked rapidly, wiping her eyes with the corner of her apron.

"I don't know how to stop feeling," Adrienna

finally whispered.

Kerilee softened. "You don't have to. You just have to stop letting it rule you."

The uneasy stillness broke as John Fisher came sprinting through the trees, his tricorn hat clutched in one hand. His face, usually so composed, was drawn tight with urgency.

"They're coming!" he shouted as he neared. "The British are marching on the town. They're closer than we thought!"

Kerilee stiffened. "How much time do we have?"

John didn't waste breath on a long explanation. "An hour. Maybe less. The locals are trying to evacuate, but there aren't enough horses. Not for everyone."

Mysty, crouching atop a branch overhead, peered down with a grimace. "Any chance the British are polite time-travel enthusiasts?"

John shot her a flat look. "None whatsoever."

Already, faint vibrations trembled through the ground. The distant clomp of hooves and the low rumble of cannon wheels echoed across the hills, growing steadily louder. Between the trees, flashes of scarlet uniforms flickered like drops of blood on the move, their muskets catching the sun in lethal flashes.

"We need to move," Roger said sharply, glancing toward the road leading into town. "If we can get to the woods on the far side, maybe we can lose them."

Joanna grabbed his hand instinctively, while Adrienna stiffened but said nothing.

Kerilee's heart beat rapidly. She knew enough about history to recognize the desperation in John's voice—this wasn't the kind of moment where heroes pulled off clever escapes. This was survival against overwhelming odds.

John turned, pointing toward the rising smoke columns staining the sky. "We don't have enough time. My approximation was off."

A deep boom rolled across the hills, rattling leaves from branches as cannon fire cracked the air. Another blast followed—closer, shaking the ground beneath their feet.

John's voice was grim, final. "The town's already captured. There is no escape now."

Adrienna's fingers found Roger's hand and squeezed tight.

"Then we stand and fight," she said fiercely, her eyes flashing.

"Well, we're doomed," Mysty muttered.

# CHAPTER 20: LOCKED IN A BARN

The heavy smell of hay, sweat, and damp wood pressed down on Kerilee as she stirred awake. Dim light filtered through cracks in the barn walls. Above her, thick beams sagged under their own weight. Beneath her, scratchy hay poked through her clothes.

Roger slumped against a beam nearby, arms loosely around a shivering Joanna. Adrienna paced the narrow loft space, her boots grinding softly in the dust. Mysty perched at the loft's edge, ears flicking with every sound from below.

Kerilee sat up, wincing. Every muscle ached. The clatter of metal and the low bark of British soldiers echoed faintly beyond the barn doors.

"At least we're alive," Roger rasped.

"For now," Adrienna muttered, shooting a wary glance at the thick doors. "They could change their minds."

Kerilee hugged her knees, mind racing. Why hadn't they been executed on the spot? Prisoners had no guarantees—especially prisoners dressed strangely.

Joanna whimpered and pressed closer to Roger, her fingers knotting into her borrowed dress. Adrienna's pacing quickened, agitation bleeding from every step.

"They think we're witches," Mysty said dryly, voice barely louder than rustling hay. "Half of them are too superstitious to touch us. The other half are just waiting for an excuse."

Kerilee's stomach twisted. Waiting for disaster was worse than facing it. She scanned the loft—the hay chute was blocked with crates, the ladder below guarded.

"We need a plan," she whispered.

Mysty's ears twitched toward the barn door. "I'm working on it."

Roger shifted against the beam, grimacing. He glanced at Kerilee and Adrienna, face drawn.

"I should probably tell you something," he said quietly. "Before I ended up here... I woke up somewhere else."

Adrienna paused mid-pace, narrowing her eyes. "Where?"

"Upstate New York, a second time," Roger said. "But not now. It was the late 1820s. John Fisher was there too—helping a minister start a church."

Kerilee leaned in, every nerve alert. "You traveled forward before you traveled back?"

Roger gave a short, humorless laugh. "Apparently. Time doesn't seem to care about direction."

Adrienna crossed her arms. "And how exactly did you get back here?"

Roger shrugged helplessly. "I fell asleep."

From her spot against Roger, Joanna lifted her head. "It's the same for me. Except I never arrive at the same time or place as Roger. Always a little after. Like I'm chasing him."

Her voice wavered. She dropped her gaze.

"And where did you end up when you got here?" Kerilee asked gently.

Joanna smoothed her skirt nervously. "Inside someone's house. A kind maid found me, gave me these clothes before the soldiers came."

Adrienna muttered, "Nice to know someone's having a softer landing."

Kerilee tucked the information away. Sleep, proximity, emotion... Maybe even dreams themselves were tethered to the curse. If they didn't figure it out soon, they might never get home.

Rough laughter floated up from the stables below. The sharp, ugly sound twisted Kerilee's gut.

"Three witches upstairs," one soldier jeered. "Maybe we should see if they float in the river later!"

Another barked a crude joke that made Joanna shrink tighter against Roger. Adrienna's jaw clenched hard enough that they made a popping sound.

She leaned toward the loft's edge and muttered, "Sorry, boys, not on the menu."

Kerilee stifled a nervous laugh. "We need out of here," she whispered. "Before curiosity turns to cruelty."

Mysty flicked her tail once, ears cocked toward the door. "Working on it," she murmured low enough that only Kerilee heard.

The tension in the loft pulled tighter. Every stomp of boots and rattle of muskets below sent Kerilee's heart hammering harder. They were trapped—and time was running out.

Mysty slinked toward the loft's edge, sleek and low. Pigs snuffled lazily in a pen. Horses shifted in their stalls. A few chickens scratched at the dust. A fat gray mouse poked its nose from a feed sack.

Mysty's whiskers twitched. She narrowed her eyes, focusing.

With a soft, deliberate mrrrow, she sent out a ripple of intent—silent, but ancient. One by one, the animals paused, turning toward her.

The proud black stallion pawed the ground, tossing his mane. He understood.

A pig near the door snorted and shuffled forward. Even the mouse twitched its nose in a tiny nod.

"If the redcoats open the door," Mysty whispered, "we'll create a racket. You all escape out the hay chute."

Kerilee blinked, glancing between Mysty and the nervous horses. "Are you sure?"

Mysty's tail flicked. "They want out as badly as we do. Give them the chance—they'll create enough chaos for a small army."

Kerilee relayed the plan to Roger, Adrienna, and Joanna.

Adrienna's mouth twitched into a rare, grim smile. "We're trusting a cat to run our jailbreak?"

Roger muttered, "At this point, I'd trust a talking chicken if it got us out."

They crouched near the loft's edge, every breath held.

The barn door creaked.

Time to move.

A cold draft rushed in as the barn doors swung wide, carrying the stink of gunpowder and damp wool with it. Three British soldiers stomped in, muskets ready, bayonets glinting.

"You, three girls," the tallest barked, motioning to the loft. "With us."

Roger started to rise, but Adrienna yanked him down.

Before anyone could move, Mysty let out a sharp chirrup—the signal.

The black stallion screamed, rearing high.

Hooves slammed into the stall door, splintering wood. Pigs squealed and bolted. Chickens erupted in a flurry of wings and screeches, pecking at arms and faces.

Kerilee's pulse thundered. Joanna's eyes widened, her whole body locking up as screams and musket fire exploded around them. For one awful second, she stood frozen—eyes wide, limbs locked.

One soldier fired wildly into the ceiling, raining hay and dust.

"Now!" Mysty yowled.

Kerilee, Roger, Adrienna, and Joanna moved as one. Forget the hay chute—they sprinted for the side ladder. In the chaos, a soldier dropped his musket. Roger snatched it up.

Adrienna kicked a feed bucket off the ledge, sending it clanging into a soldier and knocking him sprawling.

Kerilee grabbed Joanna's hand. "Move, now!"

Joanna hesitated—but Roger tugged her along.

The side door hung open. Without looking back, they burst into the fields, the cold air hitting like a slap.

They didn't stop until the barn was a smoky smear behind them. The fields bled into a rocky trail, half-choked with brambles.

Kerilee gasped for breath, legs burning. Roger led, Joanna close behind. Adrienna covered the rear, musket gripped tight.

Mysty bounded ahead, a blur of black fur weaving through the scrub.

At last, the trees thinned—and campfires flickered ahead. Tents sprawled across a clearing, canvas sides snapping in the wind. The sharp smells of woodsmoke and boiled stew filled the air.

"A grizzled officer stepped into their path, musket raised, his jaw bristled with gray stubble, eyes sharp beneath bushy brows. His coat hung loose, missing buttons, the sash at his waist stained and fraying."

"Halt!" he barked, scanning them sharply. "Drop those."

Roger obeyed. Adrienna hesitated, then dropped her musket with a huff.

"You're no redcoats," the officer muttered. "Too ragged by half. Get them fed, clothed, and tented!" he shouted to the camp.

Soldiers hurried forward to escort them.

Roger sagged, slinging an arm around Joanna. "We made it," he breathed.

Kerilee stumbled toward a fire, accepting a tin cup of soup from a kind hand. She dropped to the ground with a grateful sigh.

Adrienna sat beside her, scanning the camp with wary eyes. Joanna clung to Roger, pale but determined.

Roger glanced at Adrienna with a crooked grin. "Not bad for a witch jailbreak, huh?"

For the first time, Kerilee dared to believe they might be safe—at least for a little while.

She lifted the cup to her lips—

A searing pain knifed through her skull.

"Kerilee gasped, clutching her head as the world smeared—firelight stretched, voices melted into echoes, the camp twisting like a dream unraveling."

The world spun, and for a moment Kerilee heard a distant, distorted chant—low, wordless, like the earth itself was humming. The air around her blurred, colors bleeding at the edges as the vision swallowed her whole.

In a vision, she saw herself, Adrienna, Joanna, and Roger—standing solemnly in a strange courtyard or garden. Each, trembling, placed a pinch of dirt into their mouths. Their faces were grim. Resigned.

Kerilee snapped back with a choked sob, phantom soil clinging to her tongue.

Adrienna gripped her arm. "Kerilee! What was that?"

Kerilee shivered violently.

Her voice came ragged and low. "It's what comes next."

# CHAPTER 21: RETURN OF TIP AND EBERHARD

Karyn Molen wiped her hands on a faded towel, the Windmill Diner buzzing with low conversation and clinking silverware. The scent of fried chicken, biscuits, and sweet potato pie filled the air, comforting and familiar.

The bell over the door jingled.

Karyn's gaze snapped toward the entrance—and there they were. Eberhard and Tip, casual as anything, stepping into the diner like they hadn't been missing for days. Eberhard's long coat brushed the floor behind him, his light brown hair catching the evening light. Tip, in a crumpled flannel and jeans, whistled low under his breath as he looked around for an empty spot.

They moved toward their usual booth without a word. A few regulars glanced up but quickly returned to their meals; strange as they were, Eberhard and Tip were part of the town's landscape.

Karyn snatched up her order pad, feeling a dozen questions pressing against her lips, but held them back for now.

Eberhard slid into his seat. Tip flopped opposite him, already studying the menu even though he ordered the same thing every time.

"Fried chicken," Tip said brightly when Karyn approached. "Double plates. Extra biscuits. Oh, and hot cocoa so strong it could wake Rip Van Winkle himself."

Eberhard gave a slight nod of agreement. "Lemonade for me. Fresh squeezed. Not powder."

Karyn braced her hands on her hips, narrowing her gaze at them both "Where in Molenhaven have you two been?"

Tip grinned wide enough to flash biscuit

crumbs left over from his last meal. Eberhard only gave her a small, unreadable smile as he unfolded his napkin with deliberate care.

Karyn tapped her pen impatiently against her pad, but neither seemed in any hurry to answer.

Tip took a giant gulp of the hot cocoa when it arrived, wiped his mouth on his sleeve, and leaned back like a man ready to spin a tall tale.

"We went to see Eberhard's family," he said, grinning around the rim of his cup.

Karyn blinked. She glanced at Eberhard, half-expecting him to laugh or correct Tip—but Eberhard only gave a slow nod.

"You... have a family?" Karyn asked, skeptical.

Eberhard folded his hands neatly on the table. "I do. We gathered for a reunion. It's rare. Very rare."

Karyn opened her mouth, then closed it again. Rare, sure—but with Eberhard, rare could mean once every hundred years.

Tip shoveled half a biscuit into his mouth and added, "Whole lotta folks who don't look a thing like each other. Tall, short, old, young... every hair color you can imagine. Strange bunch, but somehow they're all related."

Karyn narrowed her eyes, suspicion prickling at her gut. She knew better than to press Eberhard when he decided to be cryptic, but that didn't mean she liked it.

"You better not have picked up anything weird while you were gone," she said, only half-joking.

Tip winked. "Define weird."

Eberhard's mouth quirked just slightly at the corner—almost a smile.

Karyn slipped her pad into her apron pocket and leaned on the edge of the booth, lowering her voice.

"While you two were off playing mystery family," she said, "things went sideways here."

Tip froze mid-bite, a biscuit half-raised to his mouth. Eberhard's hands remained calmly folded, but his pale eyes sharpened.

"Ruth Embrey's been abducted," Karyn continued, her voice tight. "Sylvester's out searching the woods right now with half the town behind him."

Tip's smile faded. "Ruth. Oh no."

"And that's not all," Karyn pressed on. "Roger's lost in time. Adrienna, Kerilee, and Mysty are holed up sleeping in the backroom right now—

been worn thin chasing after him."

For a heartbeat, the diner's usual clatter and hum seemed to fade under the weight of her words.

Eberhard finally lifted his lemonade, sipping it in silence, his expression unreadable.

Karyn crossed her arms. "Bad timing for a reunion, don't you think?"

Tip cleared his throat awkwardly and glanced at Eberhard, as if hoping for a translation.

Eberhard merely set his glass down with a quiet clink. "It seems," he said calmly, "we returned just in time."

At that moment, the lights above them flickered.

Once.

Twice.

The jukebox near the front counter sputtered to life, blaring a warped, off-key version of an old swing tune. The cheerful melody bent and twisted in the air, sounding almost like a dirge sung through broken glass.

Tip dropped both chicken legs onto his plate and half-rose from the booth, eyes wide. Customers jerked in their seats, heads whipping around in alarm. Silverware clattered onto plates. Someone gasped near the back.

Even Karyn, who didn't scare easily, felt her skin prickle with a sudden chill that had nothing to do with the air-conditioning.

Only Eberhard remained still, a slow breath passing between his lips.

With a single lift of his hand—no more than a tired wave—the lights steadied, the jukebox sputtered into silence, and the heavy pressure weighing on the diner evaporated like mist in the sun.

Conversation slowly resumed, hushed and jittery.

An elderly couple hurriedly paid their bill and shuffled out the door, casting nervous glances over their shoulders. A teenager near the counter dropped his milkshake with a loud splat, earning a sharp look from his mother as she hurried to clean it up. At the back of the diner, old Earl Preston muttered loud enough for everyone to hear, "When the lights flicker in Molenhaven, bad things follow. Always have."

Karyn offered a forced smile to the remaining patrons. "Just a little power surge, folks. Nothing to worry about."

Nobody looked reassured. Conversations

resumed, but quieter—like even words felt dangerous.

Tip sat back down, wiping his sweaty palms on his jeans.

"What... what the heck was that?" he muttered.

Karyn didn't answer. She didn't need to. She caught Eberhard's eye across the table—and in that moment, she understood: Mahtantu already knew they were planning to act.

And he was warning them.

Eberhard's expression darkened, the easy air he'd worn when he entered the diner now gone. He stared into the steam rising from his mug like he could see the battle ahead already unfolding.

"Mahtantu won't wait quietly. He'll always strike first to thwart any plan that is devised."

Karyn blew out a breath and rubbed the back of her neck. "If the whole town helps? If we get everybody ready?"

Eberhard's gaze lifted to hers—steady, unflinching. "Even then."

Tip, for once, stopped chewing.

"But," Eberhard added, "we'll have a fighting chance if everyone is prepared. If everyone comes back."

Karyn felt the weight of it settle deep in her chest.

Roger. Adrienna. Kerilee. Mysty. Joanna. Ruth. Each a thread—and if even one snapped, everything would unravel.

Eberhard turned his head slightly, staring out the diner window toward the darkening street, his reflection faintly merging with the shadows outside.

"If they don't all return in time," he said quietly, "nothing we do will matter."

The bell above the diner door jingled softly as another customer entered, but Karyn barely noticed.

The clock was ticking, faster than ever—and now the whole town's fate was balanced on a handful of missing souls wandering through time.

Karyn swallowed hard, remembering her mother's old words: "When the lights flicker, lock the doors." But tonight, there weren't enough locks in the world.

# CHAPTER 22: THE SECOND STOP IN TIME

Roger hit the cobblestones with a grunt. The sharp chill of the air punched into his lungs as he rolled onto his back, blinking up at a sky the color of cold iron. The air smelled of smoke, damp stone, and something sharp and herbal he couldn't place.

Groaning, he pushed himself up to sit. Around him, tall stone buildings loomed—gray and stern—with narrow windows and slanted roofs. Brown-robed figures moved through the courtyard beyond, sandals slapping the stones, speaking in a harsh, guttural language he didn't understand.

Roger staggered to his feet, brushing dust from his jeans. A few of the passing monks—or whoever they were—gave him curious glances, but none stopped. He caught fragments of their conversation, strange words that sounded like German but twisted somehow, like hearing a song played backward.

"Where am I?" he muttered, his voice sounding small in the open space.

A familiar figure caught his eye across the courtyard. He'd know that sharp profile and narrowed gaze anywhere—John Fisher, no question. Except here, he wore a simple brown robe like the others and spoke animatedly with an older man in a scholar's cap. Roger's chest loosened. Finally, someone he knew.

He jogged forward. "John! Hey! John!" he called, waving an arm.

Neither man reacted. John's head remained bowed in conversation, gesturing with one hand toward a courtyard garden.

Roger closed the distance and shouted again,

louder this time. "John, it's me! Roger!"

At last, John turned. His expression shifted from deep thought to mild confusion as he squinted at Roger. He said something—again in that strange, musical German.

Roger blinked, bewildered. "What? John, it's me—Roger Van Winkle!"

John stepped closer, tilting his head. He rattled off another string of foreign syllables.

Roger's frustration boiled over. "Are you kidding me? English! You know English!"

But John just shook his head slowly, then beckoned him sharply, pointing toward the side of the courtyard where vines twisted up a crumbling stone wall.

With no better option, Roger followed.

He didn't notice the way the robed figures in the courtyard slowed to watch him go—the strange outsider in alien clothes—or the way the heavy iron bells above the courtyard gave a low, ominous groan as if stirred by his arrival.

Roger barely kept pace as John—or Johannes, as some apparently called him as they passed by—strode across the cracked flagstones toward the garden wall. Every instinct screamed at Roger to demand answers, but the language barrier hung between them like a heavy iron gate.

Johannes stopped beside a ragged patch of herbs and mossy stones. Without hesitation, he bent down, scooped up a handful of dirt, and packed it tight in one weathered palm.

Roger frowned. "What are you doing? Look, I don't need landscaping tips—I need—"

Johannes shoved the dirt straight into Roger's mouth.

Roger gagged, sputtering, almost falling over backward. He clawed at his tongue, spitting and coughing. The dirt tasted like bitter iron and wet decay, coating the inside of his cheeks with grit. He staggered a few steps away, retching into a patch of grass.

"What the—!" he choked out between coughs.

But something strange was happening.

The muttered conversations in the courtyard, the distant arguments beyond the cloisters—all of it shifted in his ears. What had sounded like thick, garbled nonsense now slid into recognizable shapes. Half-formed German turned into fractured English.

Roger wiped his mouth with the back of his sleeve, glaring at Johannes. "What the heck was that for?"

Johannes smiled faintly—his first expression of real amusement since Roger had arrived. "The language of the land lives in the earth," Johannes said, his voice now clear and accented, but understandable. "It is the only way to learn it fast. To survive."

Roger spat again for good measure, grimacing. "You could've warned me, you know."

Johannes only shrugged, as if forcing strangers to eat dirt was everyday behavior.

The robed man's eyes flickered over Roger's clothing with a faint crease of worry. "You cannot remain as you are. Your garments mark you as... other."

"Gee, thanks," Roger muttered, brushing at his jeans again. "I'm from the future and you told us to come find you."

"Today of all days," Johannes jerked his head toward a narrow alley between two buildings. "Hide there. I will find you robes."

Roger opened his mouth to argue—but the wary glances from the courtyard monks silenced him. Grumbling under his breath, he darted into the shadows, pressing himself against the cold stone wall.

He watched Johannes stride away, his brown robe blending into the throng, until he disappeared into the monastery's side entrance.

Alone now, Roger rubbed the back of his neck and muttered, "Eating dirt, wearing robes... What's next? Sheep herding?"

The iron bells above gave another low groan, as if answering.

Roger shivered.

Something about this place didn't just feel old—it felt alive, and it was watching.

Roger fidgeted in the alleyway, shifting from foot to foot as the cold stone leached warmth from his back. He squinted toward the monastery door where Johannes had disappeared, but there was no sign of him returning.

A noise to his left made him tense. Footsteps—light but quick—echoed against the walls. Roger crouched, half-ready to bolt if needed.

"Roger!" a voice hissed urgently.

His heart leapt. He knew that voice.

Peering around the corner, he caught sight of a figure rounding the building—a young woman in a simple blue dress, her hair pulled back in hurried braids.

Joanna.

Relief flooded him so fast he nearly stumbled as he rushed to meet her. "Joanna!" he cried.

But when she spoke, her words were an incomprehensible tangle—sharp and strange, like the twisted German he'd heard before the dirt trick. Joanna gestured wildly, her brow furrowed in frustration, clearly trying to communicate but failing miserably.

Roger reached down instinctively, scooping a handful of dirt from the edge of the alley.

Joanna saw what he was about to do and smacked his hand hard enough to send the dirt flying.

"Hey!" Roger yelped, cradling his wrist. "What was that for?"

Joanna pointed at her mouth, then shook her head vigorously. Her hands flew in a series of gestures Roger couldn't quite follow, but he got the gist: *Don't eat that!*

Before he could argue, more movement caught his eye.

At the monastery's side gate, three familiar figures appeared—Kerilee, Adrienna, and Mysty.

Roger's heart jumped again. He waved frantically, calling out, "Over here! Thank goodness!"

The women rushed toward him—Kerilee in the lead, Mysty perched on her shoulder, looking equally annoyed and alert.

But as they reached him, confusion flickered across their faces. Mysty's fur bristled. Adrienna's mouth moved rapidly, but once again, Roger couldn't understand a word she said. It was like trying to tune in a broken radio.

Roger slapped his forehead. "Oh, come on! Again with this?"

Kerilee seemed to grasp the situation faster than the others. She bent, scooped up a pinch of dirt, and gave it a wary sniff.

Then, grimacing, she tossed it into her mouth.

Mysty's eyes widened in horror. "Mowe eh ech eh kinewoop," she muttered—though to Roger, her words were weird.

Kerilee shuddered and spat into the grass, wiping her tongue with the sleeve of her hoodie. But when she looked up, she nodded sharply to Roger. "Now I understand you. Gross, but effective."

Roger nodded back. "Trust me, not my idea."

Kerilee quickly scooped a handful of dirt and thrust it at Adrienna. "Eat it. You'll understand."

Adrienna wrinkled her nose. "Yosh cobba ne la Moch schee."

Kerilee shoved the dirt closer.

Grimacing like she was about to chug a cup of motor oil, Adrienna pinched some between two fingers and popped it into her mouth.

She gagged immediately. "Ugh! It's like licking a boot."

Mysty hopped down, grumbling. "Boosh wrigg elle ne," she muttered, and daintily flicked a few grains into her mouth with a paw.

A jolt shivered down Roger's spine as Mysty's voice snapped into clarity: "If I grow an extra tail, it's on you."

Despite everything, Roger laughed—a short, shaky sound that made Joanna frown in confusion.

Roger knelt quickly and offered Joanna another handful of cleaner dirt, scooped carefully from under a rose bush.

This time, she hesitated only a second before stuffing it in her mouth. Her face twisted in disgust, but when she swallowed, her eyes widened.

"Roger?" she said clearly, voice trembling.

Roger grinned. "Finally."

Joanna threw her arms around him, nearly knocking him off balance.

Johannes reappeared a few minutes later, a bundle of brown robes draped over one arm. His face was tense, and he wasted no time tossing the garments toward Roger and the others.

"Quickly. You must blend in."

Roger fumbled with the rough wool, pulling the robe over his head. It smelled faintly of incense smoke and old earth. Kerilee, Adrienna, and Joanna followed suit, though Mysty simply shook herself and declared, "No robe fits this much majesty," before trotting alongside Kerilee's hem.

"Come," Johannes urged, glancing nervously over his shoulder.

They slipped through a narrow passageway behind the cloisters, stone walls pressing close on either side. The air grew cooler as they descended a tight spiral staircase carved into the stone itself. The only light came from tiny slit windows set high in the walls, letting in slivers of the overcast sky.

Finally, they emerged into a small, hidden annex at the edge of the monastery—a quiet refuge away from the bustle of the courtyard. A shallow stone well sat in the center, and carved benches lined the walls. Herbs hung drying from overhead beams, filling the space with the soothing scents of lavender, rosemary, and sage.

Johannes motioned for them to rinse their mouths at the well. "The dirt leaves traces. Best to clear it before someone notices."

They obediently swished the cold, mineral-tasting water, spitting discreetly into a nearby drain.

Roger wiped his mouth on his sleeve and turned to Johannes. "Where are we exactly? Germany, right?"

Johannes nodded. "Wittenberg. In the year of our Lord fifteen hundred and seventeen."

Kerilee's jaw dropped. "Wittenberg? Wait—Martin Luther? The Reformation?"

Johannes smiled faintly. "Yes. He is here. Today, in fact, he is posting his theses."

Kerilee looked ready to faint with excitement. "We're witnessing history."

Adrienna snorted. "Great. Let's not get nailed

to anything ourselves. Oh, did I mention that I've lost that bloody nail. Oh yeah, I did. But, I still have that Willow Bark Karyn gave me."

Mysty perched on a bench, licking a paw. "Or burned at a stake. I hear they're very thorough about that."

Johannes's shocked gaze rested on Adrienna, "What bloody nail?"

Adrienna looked to Kerilee who encourage her to continue. "Joanna met us in the future where she gave us a dried blood-covered nail that she said Roger gave her."

"Ewwww, a bloody nail," Joanna whined softly with a pout.

"Where would I get a nail from?" Roger asked scratching behind his ear.

Johannes clutched at his heart. "You had the nail but it *did not* come with you. You also had Willow Bark and it *did* come with you."

"That's right," Adrienna nodded.

Mysty poked Johannes' leg. "Why is the missing nail significant?"

Johannes folded his arms. "It means that time has reset and that some things in past history could be altered now. Tell me, what did the rest of you bring?"

Kerilee shook her head, "Just the pendant you gave me."

Pointing to Roger, Johannes said, "To find him?"

Kerilee nodded.

Roger pulled out a large bottle of sleeping pills. "I brought these."

"Sleeping pills?" said Adrienna snatching the bottle from Roger's hands. "You are unresponsive in a hospital because you took sleeping pills!"

"I only took the recommended dose," Roger replied.

Johannes raised his hands, "Please calm your voices. Remember you cannot be found here associated with Martin Luther or myself."

Roger frowned. "Wait, Martin Luther—wasn't he the guy who started the Reformation?"

Mysty smacked her paw between her eyes. "You're a little slow today, Roger. We were just talking about him two minutes ago."

Kerilee nodded. "He challenged the practices of the Church. It shook Europe to its core."

Johannes studied them quietly, a shadow crossing his face. "And it will not be without blood. Not today, perhaps, yet I expect that it shall soon come."

A loud thud echoed from above, followed by shouting in angry, slashing German.

Johannes stiffened. "They have seen Martin's theses. The backlash begins."

Roger's stomach dropped. "Are we in danger?"

Johannes didn't sugarcoat it. "Only if they link you to me—or Martin."

Kerilee muttered, "I thought he was just trying to start a discussion."

Johannes gave a bitter smile. "Some discussions end in fire."

Joanna edged closer to Roger, gripping his sleeve tightly. He squeezed her hand back, wishing he could promise everything would be fine.

But the way Johannes was already glancing toward a hidden door at the far end of the annex told Roger the truth.

They weren't staying here long.

Another thud shook the rafters above, louder this time—followed by the unmistakable crash of wood splintering.

Roger flinched. "That didn't sound like a peaceful theological debate."

Kerilee tightened her robe around her and looked anxiously at Johannes. "What's happening?"

Johannes tilted his head, listening intently. His face darkened. "Some of the clergy have seen Luther's words posted publicly. They consider it heresy. They're calling for his immediate arrest."

Roger swallowed hard. "What about us?"

Johannes's eyes flicked toward the hidden door again. "You are foreigners. Out of place. Suspicious. If they catch you with me..." His voice trailed off, grim.

Above them, boots pounded along the stone floors. Angry shouts rang out—harsh accusations, demands for loyalty to the old ways. The air grew tight with fear.

Mysty hopped onto the stone well and flexed her claws. "I'm not liking the way this smells. Smells like torches and pitchforks to me."

Adrienna backed away from the wall instinctively. "We have to get out of here. Now."

Johannes nodded sharply. "There's a tunnel. Hidden beneath the old cloisters. It leads out past the city wall to the riverbank."

Roger's heart hammered. "Then what?"

Johannes grimaced. "Find a boat. Get across the river. Head east. Far east."

Kerilee glanced at the others, gauging

their readiness. She saw the tightness around Adrienna's mouth, the stubborn tilt of Joanna's chin, the way Roger's fists clenched at his sides.

They were ready—or as ready as they would ever be.

"Lead the way," Kerilee said.

Johannes moved swiftly, pulling aside a tapestry that depicted a saint wrestling a dragon. Behind it was a rough stone door, the outline almost invisible unless you knew where to look.

He pressed a hidden latch. The door creaked open, revealing a narrow tunnel plunging into darkness.

"Stay close," Johannes warned. "No light until we're past the first bend."

The group filed inside, the air turning colder and damper with each step. Their robes whispered against the stone walls as they moved deeper underground, the distant shouts of the angry crowd growing fainter—but no less urgent.

Thunder rumbled overhead, shaking dust from the tunnel ceiling. A distant bell began to toll—a low, ominous sound that set Roger's teeth on edge.

As they turned the first corner, Mysty

muttered low enough for only Roger to hear, "I hate history. Too many mobs. Not enough common sense."

Roger couldn't help a grim smile.

Ahead, Johannes motioned for silence, pointing toward a faint sliver of light where the tunnel angled upward.

"This is where you run," Johannes whispered. "Follow the river. Find a boat. Disappear."

"What about you?" Kerilee asked.

Johannes hesitated, his expression unreadable. "I have my part to play."

Without waiting for more questions, he pushed open a low hatch, letting in a blast of damp, fog-choked air. Beyond it, they could see the misty outline of the river, black and slow under the darkening sky.

Cold, misty air greeted them. They stumbled onto a fog-shrouded riverbank.

"There!" Mysty pointed with her nose.

A small wooden boat bobbed in the mist.

Roger didn't hesitate. He helped Joanna in first, then turned to the others. "Move!"

Kerilee and Adrienna clambered aboard. Thunder cracked overhead.

Behind them, flickering torches swarmed

through the city's narrow streets.

"They're coming," Kerilee said grimly, grabbing an oar.

Roger pulled Joanna closer as the boat drifted into fog. "We'll make it," he promised softly—even if he wasn't sure.

Roger shoved the boat free. It slipped into the sluggish current.

Behind them, the tolling bells of Wittenberg grew frantic, echoing into the mist—marking not just the passing of time, but the spark of a firestorm that would reshape the world.

# CHAPTER 23: THE STALKER

Ruth Embrey woke to the sting of blood dried at the corner of her mouth and the sharp ache of her wrists bound tightly behind her. Her black hair hung loose and wild in her face, clinging to sweat and grime.

She sat on the dirt floor of a room in shambles, her back pressed against a splintered support beam. Rough rope bit into her skin. Broken boards and scraps of plaster littered the ground around her.

Across the room, Lester and Cobb Dimwitty

loomed, their faces twisted with glee and something darker. Lester held a jagged object that glowed faintly with sickly green light. Cobb swung a heavy mallet in lazy arcs, testing its weight, making Ruth flinch each time it passed too close.

"Well, well," Lester drawled, scratching his scraggly beard. "Sleeping Beauty's finally awake."

Cobb chuckled as he kicked her in the back, sending a jolt of pain up Ruth's spine.

Ruth kept her face neutral despite the panic rising in her chest. She licked her cracked lips, tasting copper, and forced a smirk. "Nice welcoming committee. Missing the tea and biscuits though."

Lester's face twisted. Cobb just laughed—a low, stupid sound that made Ruth's skin crawl.

She shifted her shoulders, feigning casualness. "Where am I? Dimwitty Family Resort and Spa?"

Cobb sneered. "You ain't goin' nowhere, sweetheart."

Lester leaned in close, his breath sour. "You poked your nose too close to the Sky Witch's business."

Ruth's heart skipped, but outwardly, she rolled her eyes. "Oh no. Not the scary Sky Witch. Should

I be trembling?"

Their laughter had a hard, mean edge.

Cobb cracked his knuckles. "Think you're real funny, huh?"

Ruth smiled coldly. "Funnier than you two playing henchmen for a third-rate demon."

Their grins faltered—and Ruth's pulse quickened. Good. She still had some power left.

Cobb stepped forward, raising a hand.

Ruth didn't flinch.

She wasn't giving them that satisfaction.

The slap came fast and sharp. Lester's hand cracked across Ruth's face, snapping her head sideways.

Pain exploded through her cheekbone. Her lip split open fresh. Warm blood slid down her chin.

Still, Ruth refused to cry out. She turned her head back slowly, fixing them with a glare that could melt steel.

"Is that all you got?" she rasped.

Cobb's nostrils flared. Lester shook his stinging hand and barked, "What did she tell you?"

Ruth's voice was icy. "That Mahtantu's days are numbered."

Their faces twisted in unison—shock turning

to fury.

"You little witch," Cobb snarled, swinging the mallet menacingly.

"She's gotta know something," Lester hissed. "Why else would she poke her nose where it don't belong?"

Lester raised the glowing object between them, its sickly green light pulsing brighter, casting warped shadows across the walls. The air tasted faintly metallic, like a brewing storm.

Cobb's eyes widened. "Don't—Mahtantu said to save that for the Sky Woman."

Lester hesitated, fingers twitching around the jagged shape. "Fine," he muttered, pulling it back—but the glow lingered in the dark like a warning.

Cobb swung the mallet hard into Ruth's shin.

The agony made her see stars. A muffled scream ripped from her throat before she could stop it.

Tears stung her eyes—but Ruth forced herself to breathe slow and steady through the blinding pain. No begging. No breaking.

Cobb grinned at the pained sound. "Tough girl, huh? Bet you won't be so mouthy now."

Ruth spat blood onto the floor at their feet.

"You're still ugly," she muttered.

Cobb raised the mallet again, but Lester grabbed his arm. "Not yet. Mahtantu might want her alive."

"Yeah, I thought you said Mahtantu was going to make me forget?"

Lester shot Cobb a glare. "You shouldn't believe everything Cobb says."

They snarled at each other, torn between wanting to hurt her and fearing Mahtantu's wrath.

Finally, Cobb ripped a strip of duct tape from a roll and slapped it over Ruth's mouth. She tasted adhesive and blood, gagging slightly but refusing to struggle.

"You stew in that," Cobb said, giving her a rough shove that knocked her sideways against the beam.

They turned away, arguing in low, furious voices.

Ruth slumped against the beam, every bruise throbbing, her breath shallow. Her vision blurred at the edges, but she refused to let the darkness win. Her pulse beat stubborn and slow in her ears.

The door burst inward with a deafening crash.

A figure hurled itself through the opening—tall, broad-shouldered, fists already swinging.

Orrin Sutton hit Lester first, a brutal punch that snapped his head sideways and dropped him. Before Cobb could react, Orrin pivoted and slammed a heavy boot into his gut. Cobb folded with a grunt.

Breathing hard, Orrin spun to Ruth.

"Hang on, Ruthie," he muttered, pulling a hunting knife from his belt. With a few swift slashes, the ropes fell away.

Ruth peeled the duct tape from her mouth with shaking hands, coughing and spitting the foul taste onto the floor.

"How—" she rasped, "How did you find me?"

Orrin flashed a grim smile. "You really think I'm the kind of guy who'd let you out of my sight for long?"

Ruth blinked up at him, pain pounding in her shin. She wasn't sure whether to be furious or relieved—or both.

Ruth leaned heavily against Orrin. He wrapped an arm around her back carefully, steadying her without making her feel weak.

"You're limping bad," he said, voice low.

"No kidding," Ruth muttered, grimacing.

His hand hovered uncertainly over her side.

"Don't even think about carrying me," she warned.

"You're stubborn," Orrin said, smiling despite himself.

"And you're a terrible stalker," Ruth shot back. "You let me get kidnapped."

Orrin chuckled. "I wasn't stalking-stalking. More like... keeping an eye on you."

"That's supposed to sound better?"

He shrugged. "I knew the Dimwittys were getting bolder."

"Well, great job there, Sutton." She tried to smirk but winced. "Maybe next time, stalk harder."

A low laugh rumbled from him.

"Truth is," Ruth said, softening, "I might've been... keeping an eye on you too."

Orrin blinked. "Seriously?"

"Seriously. Didn't expect you to turn out so...heroic."

He lifted her more fully into his arms, careful of her injuries. "You always were the brave one."

Ruth rolled her eyes but leaned into him. "Yeah, well. About time someone noticed."

Orrin half-carried Ruth out of the crumbling

house. The cool slap of night air hit her face as they stumbled onto the porch.

A window shattered behind them.

"Get to the truck!" Orrin barked, tightening his grip.

Through the corner of her eye, Ruth saw Lester and Cobb stumble out with rifles.

A shot rang out.

The rear window of Orrin's beat-up pickup exploded in a spray of glass shards.

He threw open the passenger door, shoved her inside, and slammed it. Another bullet pinged off the side mirror.

Orrin dove into the driver's seat, jamming the keys into the ignition.

The engine roared to life.

"They're following us!" Ruth shouted, twisting to look back.

Orrin grinned fiercely. "Not for long. I slashed their tires."

The tires screeched as he floored the gas. The truck fishtailed wildly before gripping the dirt road.

Behind them, furious shouting faded—punctuated by desperate gunfire.

Trees blurred past. The engine growled. Ruth

pressed a trembling hand against her bruised side, heart pounding.

Orrin glanced at her.

"You alright?" he asked.

Ruth smiled grimly. "Yeah. Thanks to you, Stalker."

He laughed—a short, raw sound—and pushed the truck faster into the night.

But as the truck tore down the forest road, Ruth caught movement in the tree line —shadowy figures slipping between the trees, keeping pace.

A chill ran down Ruth's spine. Whatever moved alongside them wasn't entirely human.

Her heart hammered harder.

"Orrin," she whispered, "I don't think it's just them."

The Dimwittys weren't alone.

"They're not giving up," Ruth rasped.

Orrin's hands tightened on the wheel, jaw locking.

"Then neither are we," he growled, as the forest seemed to close in around them.

# CHAPTER 24: THIRD STOP IN TIME

Kerilee stumbled forward, boots scraping over cold stone. The air smelled damp and smoky, and the flickering torchlight that lined the corridor cast long, broken shadows against the ancient walls. Mysty yowled lowly from somewhere near her ankles. Adrienna muttered a curse under her breath, while Joanna gasped behind them.

"Is it just me," Adrienna whispered, "or is it way too quiet for a place this creepy?"

Kerilee braced a hand against the wall, stone biting cold into her palm. "Where's Roger?" she hissed, panic rising up her spine.

They'd fallen through time together—but Roger had vanished. Only the echo of their own breathing and the distant crackle of flames filled the air.

Adrienna sniffed. "Hey... anyone else smell soup?"

A cold knot coiled in Kerilee's belly. "Yeah. And I don't think it's dinner."

They crept down the corridor, each step kicking up motes of ancient dust. Ahead, a faint rectangle of light marked a small arched window set into the stone. Kerilee reached it first, careful to stay low, and peered outside.

Beneath them sprawled a wide courtyard, firelight flickering off stone walls like restless shadows. Hooded figures milled about, some shouting in Latin, their voices sharp and eager. In the center of the gathering, a massive black cauldron bubbled ominously.

Kerilee's stomach turned.

"That's not soup," she muttered.

Mysty's tail puffed up like a bottlebrush. 'Definitely not soup,' she growled.

The scene below rippled with dread, and Kerilee knew without being told—they had landed somewhere very, very dangerous.

The crowd below roared as soldiers dragged a man forward, chains clanking around his ankles and wrists. His head hung low, hair matted with sweat and grime. Two guards roughly hauled him toward the cauldron, its thick black liquid hissing and bubbling like something alive.

Kerilee recoiled instinctively from the window, but forced herself to keep watching. Beside her, Adrienna covered her mouth with a trembling hand. Joanna clutched the edge of the windowsill so tightly her knuckles whitened.

Mysty growled again, the sound deep and vibrating in her throat. "Something's wrong," she said, fur bristling.

The man stumbled to his knees. A guard yanked him upright by the hair—and for a fleeting second, his face turned toward the window.

Joanna sucked in a sharp breath. "That's John Fisher!" she gasped.

Kerilee blinked, her pulse hammered in her throat. It was John—only younger, roughened by captivity but unmistakably him.

The guards shouted something in Latin, and the crowd responded with a vicious cheer. Without ceremony, they heaved John Fisher into the vat of boiling oil.

Kerilee's breath caught.

But instead of screaming, John simply sat down into the bubbling liquid—calm, still, his face blank and serene.

The shouts of glee faltered. Confused murmurs rippled through the crowd.

The boiling oil churned and spat around him, yet John didn't flinch. He sat there like a stone amid a storm, eyes lifted toward the darkened sky.

The roaring crowd began shifting backward, fear infecting the air.

Kerilee pressed a palm to the rough wall to steady herself. The chill of stone seeped through her skin. 'Where's Roger?' she hissed, panic curling up her spine. She had seen many strange things since arriving in Molenhaven. But nothing like this.

Not even close.

Mysty's ears twitched sharply. She turned from the window, nose in the air, whiskers vibrating.

"He's here," she said in a low, urgent voice.

Kerilee twisted around, heart still hammering, and spotted a shadow moving along the wall behind them. Before she could call out, a hand clamped down gently on her shoulder.

"Don't scream," came Roger's familiar whisper.

Kerilee nearly sagged in relief. Adrienna let out a choked laugh and threw her arms around him in a fierce hug. Joanna wiped at her eyes, a shaky smile breaking through her shock.

Roger crouched beside them, his face pale but determined. "You saw that?" he whispered, glancing toward the courtyard.

Kerilee nodded mutely. Adrienna shuddered. "Saw it. Definitely didn't want to see it."

"He's been doing that since I arrived," Roger said, voice low and tense. "They try to kill him every few days. Boiling oil, fire, hanging—it doesn't matter. He just… sits there. Then he gets up and walks away."

Kerilee stared at him. "How long have you been here?"

Roger scratched the back of his neck. "Long enough to realize they're terrified of him. Long enough to realize—" He hesitated, then said, "—that John Fisher isn't just some wandering scholar."

Adrienna muttered, "Anyone else want to rethink who John really is? Because I'm about ready to throw my original guesses out the window."

Kerilee didn't answer. Her mind was too busy racing, fitting puzzle pieces together that she hadn't even known were missing.

And every new piece pointed to one undeniable fact:

John Fisher was far more than he appeared.

John climbed slowly out of the cauldron, chains gone, thick black oil dripping from his robes and skin like rainwater. Where the liquid should have burned and scalded, it simply slid off him, leaving his flesh untouched.

The crowd that had moments ago cheered now stumbled backward in fear. Shouts in Latin turned into frantic murmuring, and a few spectators even fled the courtyard, their torches bobbing like fireflies into the night.

"He's not even hurt," Joanna whispered, clutching Roger's arm.

Kerilee barely breathed as they watched John

stride calmly across the stones, oil trailing behind him like a dark river. No one dared block his path. Soldiers who had moments ago seemed eager to celebrate now lowered their weapons, crossing themselves in terror.

John's movements were deliberate, almost ritualistic. Without hesitation, he disappeared through a massive stone archway leading out of the square.

Roger nodded toward the retreating figure. "Come on. He'll clean up first. Then... he'll argue with the Gnostics."

Kerilee tore her eyes away from the scene long enough to glance sharply at him. "How do you know that?"

Roger shrugged awkwardly. "I've been following him since I landed. It's like... history unfolding. And he keeps muttering about the Gnostics."

Joanna frowned. "Wait—what are Gnostics?"

Kerilee, falling into instinctive librarian mode, whispered, "They were an early religious group. They believed in hidden knowledge and secret wisdom. They were... controversial."

Adrienna gave a short, skeptical laugh. "So basically, the ancient version of conspiracy theorists."

Kerilee shot her a look. "Not exactly. Some of their ideas were profound—and dangerous in the wrong hands."

Mysty, tail twitching, added dryly, "Judging by the cauldron stunt, I'd say 'dangerous' is putting it lightly."

Joanna stared at the archway where John had disappeared. "But... how is he still alive?"

No one had an answer.

And somehow, Kerilee had the sinking feeling that whatever John truly was, it was about to matter far more than any of them realized.

Roger led them away from the courtyard, keeping to the deep shadows cast by the stone buildings. Their footsteps were muffled by the damp ground, the air heavy with the lingering scent of smoke and fear.

"We can't stay dressed like this," Roger said, glancing at the girls' modern clothes. "If the wrong person notices, we'll end up in that cauldron—or worse."

Mysty flicked her tail. "Lovely. Nothing says vacation like being boiled alive."

They ducked into a narrow alley where a line of simple robes hung from a wooden rack outside a shop. Roger handed coins to a sleepy vendor, muttering in broken Latin. The man grunted and waved them off with a hand heavy with rings.

"Here," Roger said, thrusting low cost Roman-era clothing into each of their arms. "Put these on."

Kerilee found a place to change and slipped hers on, the rough fabric scratchy against her skin. Adrienna grumbled but followed suit, and Joanna tugged hers on awkwardly, her hair snagging in the coarse weave.

Roger crouched and scooped up a handful of dirt. "Eat it. Quick."

Adrienna grimaced. "Seriously? Again?"

Mysty flicked her tail. "If there's a worm, I'm transferring my services to another cat."

With exaggerated grimaces, the women each took a pinch and swallowed. The taste was bitter and earthy, but almost instantly, the murmured Latin drifting through the air shifted into recognizable English.

Joanna coughed and gagged. "Worst. Taste. Ever. Like it came from a litterbox."

Mysty's face tilted up, "Well, if you hear my butt say something, let me know."

Kerilee patted her back sympathetically. "At

least now we can ask for directions before we get burned at the stake."

Roger pointed across the street. "Come on. He's headed to the Roman bathhouse. That's where he always goes after... you know."

They wove through the thinning crowds, careful not to draw attention. The bathhouse loomed ahead—an ancient, steam-filled structure tucked behind towering columns.

Near the entrance, they spotted John again—this time scrubbing at his oil-soaked robes with a rough cloth, the boiling marks still faintly clinging to his skin like soot.

His face was drawn, tense—not serene like before. Something had changed.

Roger motioned for them to hang back. They watched from the shadows as John finished cleaning himself, then turned sharply, his cloak whipping around him like a dark banner.

They waited until the bathhouse emptied, then shadowed him at a careful distance as he wound through the narrow lanes.

He didn't look back as he strode into a small home nestled among the trees beyond the bathhouse—a simple stone dwelling hidden from the main road.

"We follow him, right?" Adrienna asked.

Kerilee nodded, feeling her heartbeat kick hard in her chest.. "We have to."

They crept after him, sticking close to the walls, until they reached the small house. A warm glow spilled from its single window, and the smell of baking bread and herbs filled the air.

Inside, John was already preparing a simple meal, his movements brisk and focused.

As they hesitated at the threshold, Mysty muttered, "Last chance to turn around and pretend we never saw a man bathing in lava oil."

Kerilee squared her shoulders and knocked.

John looked up as if he had been expecting them all along. His expression softened, and he motioned them inside without a word.

Once they gathered around the small hearth, Kerilee spoke first, her voice trembling slightly. "We know... about the curse. About Rip Van Winkle. About what's coming."

John set down a bowl of stew and regarded her gravely. "Then you also know... you must sleep again."

Roger stiffened. "You're sending us back?"

John's eyes gleamed with something unreadable. "Yes. I've got a lot to tell all of you. It will end your journey through time here instead of going back any further. First, we eat."

# CHAPTER 25: JOHN FISHER REVEALED

Mysty lingered at the door after dinner, her tail flicking back and forth like a windshield wiper. Her yellow eyes glinted sharply in the firelight.

"Kerilee," Mysty said solemnly, "I haven't seen a single cat since arriving here. Romans didn't eat them, did they?"

Kerilee, still reeling from everything they had seen, turned toward her. "Romans that worshipped Isis considered cats sacred. But there is no evidence I can recall that the rest of the Romans ate cats."

Mysty padded softly into the center of the room. John Fisher watched her without speaking, his face a calm, ancient mask.

"I bet they do eat cats. I've had the feeling all day that someone wanted to put me on a pizza," said Mysty.

Without hesitation, Mysty leapt onto the bench beside John, leaned in close, and whispered something into his ear. Kerilee strained to hear, but the words were lost beneath the low crackle of the fire.

John's eyebrows lifted slightly. A flicker of something—relief? dread?—crossed his face before it settled back into calm. He took a deep breath and calmly began. "About sixty or so years ago, all of you arrived in my homeland. You attempted to tell me that you were from the future. I thought you were all a little crazy at the time."

All of us broke into a grin, for we knew how crazy we must have sounded to him with every attempt to explain who we are to him in the future.

John continued, "However, myself and my

friends at the time, Kepha and Yakov, had witnessed many strange things and so we accepted you. Our eyes were slowly being opened to so many possibilities of what was possible. At that time, you knew me as Yohanon. Joanna blended in so well, she served us and became well known among us."

Joanna's face brightened.

Kerilee sat up a little taller. "John, or Yohanon, before we started back in time, you mentioned a point in which you would not be able to help Roger. You said then that only your master would be able to help him then."

Nodding with a gentle smile, John proceeded to explain, "When you found my friends and I in the past, something went very wrong with Roger. I did not know then that I would live to see your future. However, I vowed that if I ever found you all again, I would be prepared to save Roger."

Adrienna was squirming with impatience. "What was it that went very wrong?"

John's face instantly showed sorrow. "We were all gathered for supper. You, Adrienna, played a game with Kerilee and Joanna while you waited. Drinking cups were set out to be filled before being brought to the main table. You taught Kerilee and Joanna this strange game by setting one cup in the middle of you and then you each took turns bouncing an item on the table and into the cup. I remember seeing how much you three enjoyed playing this game. You had a strange bottle that contained these items."

Roger turned to Adrienna. "It was either the Willow Bark or the Sleeping Pills."

Adrienna reached into her clothing and showed him both bottles. John pointed at the larger of the bottles.

Mysty's mouth dropped open. "Oh no. The sleeping pills."

Adrienna's hand flew to her mouth.

John went on to say, "One of the women that served us took that cup from your game and set it on a tray. Kerilee tried to explain that something was in the cup, but the woman went on about her work. She filled the cups, but one she overfilled. So, she carefully lifted it and added some liquid to two other cups to make them even. Then she set them on the table. The three of you tried repeatedly to explain something to the woman whose house we were in, but she would not listen."

"I am so sorry," Adrienna whispered.

John looked upon Adrienna with kind eyes. "All is forgiven. Sometimes events were meant to happen which you never intended. It was not your game that caused what happened that night."

Kerilee noticed Joanna was breathing harder with her hands gripping her coarse clothing and staring wide-eyed at John. She touched Joanna's arm and caught fast glimpses of images running through Joanna's mind. She was being affectionate with a young man that wasn't Roger.

Being so surprised at seeing this, Kerilee released her arm and turned to John. "Did any of us form relationships in this time that you're speaking of?"

Joanna's face turned to Kerilee. "What are these strange memories I have?"

John held up both hands as he explained. "Well, yes. As I said earlier, Joanna became very close with our group. She met a young man who also followed our master. She became less attached to Roger and more attached to Saad."

Mysty thumped her tail down. "You mean he wasn't happy?"

"It wasn't that," said John. "Saad was a common name at that time and place. In time, he and Joanna were married and she became a permanent member of our group. This took Roger by surprise and one morning, he was gone. When that happened, Joanna remained while the rest of you vanished back to your normal time. But, that's the end of this story, and even though it was lamentable that we never saved Roger then, my master, whose name in my language was Isho, told me that another opportunity to save Roger would be forthcoming."

Kerilee gasped. "Isho is Aramaic for Jesus!"

John nodded. Adrienna eyes grew wide. Joanna smiled. Roger slumped against the wall.

"Oh, so you've heard of him!" said John.

Adrienna took a breath. "Of course we have." Then everyone looked at Mysty.

Mysty's head popped up. "What? Yes, I know who He is, okay?"

John shook his head. "There was this time that Mysty had a misunderstanding—

"John! John! I told you never to mention it!" Mysty practically shouted. "Cats are not perfect okay, I made a mistake. When I saw how many fish you guys brought into the boat that time, I just thought some of that must be for me, okay. Let's just drop it."

"My apologies, Mysty," said John.

There was a tense silence, but Adrienna found the perfect segue. "Was that when the Lord said he would make you and your friends fishers of men?"

"Yes," said John.

Roger then understood the obvious. "So, that's why you call yourself John Fisher."

"That's correct."

"Could the fact that we have gone back in time twice to save Roger caused atmospheric disturbances?" Kerilee asked.

"Very much so," said John.

Mysty chose that moment to stretch, "That explains the rain and thunder that stalled our departure."

"Then why don't we remember the previous attempt to save Roger?" Adrienna asked.

"Because this trip through time," John began to say, "replaces the previous attempt that failed."

"But, I still have memories," said Joanna.

Kerilee answered her before John could. "Because you remained with them. Your attachment to us was through Roger. Once you severed that, you stayed with Saad."

Joanna thought a moment. "We had children. I had a happy life in the past. Once the children had grown and Saad died, I returned to my normal time."

Joanna let go of Roger's hand. He then moved closer to Adrienna and took her hand in his.

Mysty yawned. "What did I do during all this?"

John looked at her and said simply, "You drank a lot of goat's milk."

"Sounds like something I'd do," Mysty said. "Carry on, John. I want to hear the rest of the story—without the fish part."

John exhaled and paused. "I told you the very end and the beginning because they were important to you. Now, if you all wish, I will tell you what happened in between."

"If you'll explain how you've been able to live so long and not die in boiling oil," said Kerilee.

A calm smile grew on John's face, "If I tell how I am able to live so long no matter what befalls me, would you keep my identity just among yourselves."

All answered, "Yes," even Mysty.

"Then," said John, "Let us go back to that

fateful night and the supper we were having. We had finished eating and drinking. Our master wanted to spend some time in a nearby garden and we agreed to go. Most of the followers sat in an open place there and calmly talked among themselves. While they did so, Kepha who is Peter, Yakov who is James, and I followed our master a little further into the garden. He asked us to wait while he meditated. So, we sat against a tree. There, the three of us fell asleep. Now, we have Joanna to continue the tale where we were asleep."

Joanna swallowed hard. "When the master, Isho, returned and found those three that slept, he woke Peter and said, 'Could you not watch with me one hour?' I realized then that although the names were different to me at the time, that I knew that it was the Lord speaking. These names were in the Aramaic language, but now I knew their names in English from my father's sermons in church."

Adrienna slapped a second hand over her mouth. "You mean we're the ones in the New Testament who put those apostles to sleep!"

"Yep, you're famous, Adrienna," Mysty laughed.

"Not funny, Mysty," said Kerilee. "I feel awful."

John held up a hand and everyone went silent. "All is forgiven, as I said. Your names are not mentioned and no blame is attributed to any of you."

Kerilee's brows knit together. "So, if we hadn't come back in time, would the account of you and the others falling asleep have been written?"

"Most likely it would not have been written. If you didn't do it, why would it be written?"

"Funny how time works," Roger mumbled.

"Now," said John, "for the part where you learn about me. We later went to the sea of Tiberias, which is also known as the sea of Gallilee. Our risen Master spoke to us of many things, not all of which is recorded. For the purpose of this story, I will mention one item that was partially recorded in the last chapter of the gospel that bears my name. Our master spoke to each of us about what would be expected of us thereafter. Except for me, so I followed my master to ask about myself, but Kepha, who you know as Peter, looking at me said to him, '...and what shall this man do?'"

"My master replied, 'If I will that he shall tarry till I come, what is that to thee?'"

"What does that mean exactly?" Adrienna asked.

"It means this, that I was to remain upon the earth and not taste of death *until* the Lord came again. And while I waited for that day, I would teach the truth to every creature. Not only of the gospel, but all truth. Some said that I would never die, but that is not true. Each one of us will eventually shed our mortal bodies. It is our spirit that lives on."

Kerilee looked a bit uncomfortable. It seemed that a question was forthcoming, so John looked upon her and waited. Finally, she spoke. "When we return to our proper time, how are we to interact with you?"

John smiled. "As you have been, Kerilee. If it is a matter of good and evil, call for me. If it is a matter of what is true and what is false, ask me. You are my friends and I shall always be your friend. Even when you need to make sense of the strange future events you will encounter, Kerilee. Find me and I shall give you answers."

Before Kerilee could say anything more, a strange shimmer rippled through the air—the room taking a deep, unsteady breath.

Roger stiffened. Joanna gasped. Adrienna

grabbed Kerilee's sleeve—A tingling crawled up Kerilee's spine, like standing too close to lightning. Then Adrienna's outline blurred and dissolved.

"Adrienna!" Kerilee shouted, stumbling forward. Watching her vanish felt like losing her forever—grief already settling in her chest.

Roger tried to reach for her, but his hand wavered—and then he too was gone, swallowed by the same unseen force. At least they went together.

Joanna's wide, terrified eyes locked on Kerilee's for a split second before she disappeared. Joanna had found love and now returned to her future with those wonderful memories.

Only Kerilee and Mysty remained, standing alone in the flickering firelight.

Kerilee's heart thudded, uneven and loud. She staggered a step toward the empty air—then stopped, fists curling tight at her sides.

"What's happening?" she rasped.

Mysty landed beside her, tail lashing like a whip. John waved goodbye.

"Something definitely was changed here," Mysty said, voice grim.

Kerilee swallowed hard. Fear clawed at her—

but somewhere deeper, stubborn fire answered back.

Their mission in the past was complete.

But the future still required much to be accomplished.

# CHAPTER 26: NO TIME LIKE THE PRESENT

Karyn laughs at Eberhard

Kerilee blinked against the dim light, the squeak of old diner stools and the scent of frying bacon tugging her back to consciousness. For a moment, disoriented, she wondered if she was still trapped in one of John Fisher's impossible timelines. But no—she was lying on the hard floor of the Windmill Diner's back room, a floury circle still faintly marking the floor around her.

Beside her, Adrienna stirred with a groan, pressing her palms to her temples. Mysty stretched luxuriously, her black fur puffed out, and yawned wide enough to show a full set of gleaming teeth.

"Finally," Mysty muttered, her voice thick with irritation and relief.

Kerilee pushed herself up onto her elbows, her head pounding. "What happened?"

Mysty flicked an ear. "John must have had a time limit," Mysty explained simply. "Or else we were boring him with our questions."

"I hope Roger made it back, did he?" Kerilee asked. "That revelation about Joanna should have released him."

Adrienna sat up fully, alarm sharpening her sleepy features. "Where's Roger?"

Before Kerilee could answer, a familiar, cheerful voice called from the front of the diner, "Breakfast's on! Get out here before I burn it!"

Relief flooded her so fast she almost sagged back onto the floor. She caught Adrienna's wide-eyed grin, and they scrambled upright, nearly tripping over Mysty in their rush.

"Watch the tail, ladies," Mysty said dryly,

trotting after them as they pushed through the swinging door and into the diner.

The warm, golden light of the dining area enveloped them like a hug—and there, standing behind the counter flipping pancakes, was Roger Van Winkle, looking healthier and happier than she had ever seen him.

"Roger, what did you dream about last night?" asked Adrienna.

"I dreamed about you," Roger said with a tired grin.

For the first time in what felt like forever, Kerilee dared to believe they might actually pull this off.

Roger was a whirlwind of motion behind the counter, flipping pancakes with a flourish and cracking eggs two at a time onto the griddle. The smell was mouthwatering, and the low sizzle filled the diner with the promise of something solid, something normal.

Around the largest table, faces turned to greet them—Karyn, Sylvester, Tip, Ruth, Orrin, Lucy, Phil, Jonameed, Marcus, and even Arnoldus de Groot, who gave Adrienna a gruff but unmistakably relieved nod. And sitting quietly near the window, hands folded around a steaming cup of herbal tea, was John Fisher himself, looking every bit the timeless enigma he had always been.

John's pale eyes caught Mysty's first as he walked in. He gave a small, respectful nod. "Thanks for guiding them back to get Roger."

Mysty arched her back proudly. "Just doing my job. Would've charged a fee, but the future of the world seemed more pressing."

Adrienna elbowed Kerilee lightly, smiling for the first time in what felt like days. "Look at them. They're all acting like it's a Sunday brunch."

Kerilee smiled, though a tight knot of anxiety still twisted low in her stomach. It felt good—too good. Like the calm before the worst storm imaginable.

Roger plated up a stack of pancakes and set them on the table with a wink. "Eat up. You're going to need your strength."

Kerilee slid into a seat, grateful for the food, but her mind was already racing ahead. There was still so much to do—and if Mahtantu had sensed their victory in pulling Roger back, the demon wouldn't sit quietly for long.

They had precious little time left.

Kerilee pushed her plate aside after a few polite bites, her mind buzzing louder than the diner's old ceiling fans.

"We don't have much time," she said, cutting through the chatter. The room hushed immediately, every face turning toward her.

Kerilee stood and glanced at Mysty, who gave a solemn flick of her tail, signaling go on.

"The good news is we have everything we need to lift the curse—almost. Rip Van Winkle's journal laid most of it all out. But there's a catch."

Arnoldus leaned forward, arms crossed. "Always is."

"We have to mass-produce the ancient brew," Kerilce said, "and it has to be done fast. We need to get it to every Van Winkle descendant still breathing... and to the trees at Garden Rock."

Tip rubbed a weathered hand over his face. "Some of those trees are our kin. They're still waiting on us."

Karyn nodded fiercely. "You'll have every windmill running day and night if that's what it takes. All except that black one."

"Ah, that one always been trouble!" Adrienna blurted out. Arnoldus gave her a disapproving look. Adrienna shrank in her seat. "Sorry dad. Sorry everyone."

Kerilee's eyes swept over the room. "This isn't just about a curse anymore. Mahtantu knows we're moving against him. He won't wait."

Roger, flipping another pancake onto a plate, added quietly, "If we don't act now, he'll pick us off one by one."

Kerilee met his gaze. "Exactly. We have one shot to end this. We can't afford to miss."

The weight of it settled over the room. No more illusions of safety. No more delaying.

It was all or nothing now.

Karyn was the first to break the heavy silence, tapping her pen sharply against the diner table. "Alright. First hurdle—how are we carrying the night berries without ruining them?"

Sylvester scratched his chin thoughtfully. "Sassafras baskets. Strong ones. The old ones didn't have enough bark woven in to block Mahtantu's influence. I'll fortify these. Make 'em double-thick if I have to."

Tip nodded slowly. "Don't skimp. We saw what

happened last time with LeeAnn."

A somber chill swept through the group at the memory. No one dared speak her name aloud again.

Arnoldus leaned forward, tapping the table for attention. "I've reached out to a few bottling plants across the country. They're sending jug-shaped bottles that match the old designs. We'll bottle the brew right here at the windmills."

Kerilee blinked, impressed. "You arranged that already?"

Arnoldus gave a small, rare smile. "When you see what's coming, you plan ahead."

"And what about the brewing itself?" Adrienna asked.

Karyn grinned, fierce and ready. "We'll use three of the four windmills. Crank out enough to flood the Catskills if we have to."

Mysty, perched like a tiny queen on her chair, warned, "And Mahtantu knows enough of what we're doing that he's forced to strike, because if he wants to continue his existence here, he has to destroy us."

Ruth piped up from the corner. "Eberhard left instructions. Exact timing, exact locations. All the details. We must be prepared before we begin. Then, we have one chance."

"One night," Kerilee said softly. "And if we miss it—"

"We don't miss it," Roger interrupted firmly, setting another stack of pancakes on the table.

The unspoken promise hung thick in the air.

No second chances.

Not this time.

Tip cracked his knuckles and leaned forward. "We can't do this with just the few of us. We need the whole town behind us."

Karyn nodded sharply. "I've already started spreading word. Folks who still believe in the old stories won't need much convincing. But we'll need more than belief—we'll need action."

Arnoldus pulled a crumpled list from his pocket. "I've contacted the Van Winkle descendants I could track down. Most still live within a few hours' drive. They'll come if they know it's their last shot at breaking this thing."

Sylvester scratched at his graying beard. "And the ones too stubborn to believe?"

"We'll show them proof," Kerilee said firmly. "Bring them up to Garden Rock if we have to. Let 'em see the trees for themselves."

Roger set down a fresh plate of food in front of

Karyn and muttered, "Better they're scared now than dead later."

Lucy, her braid swinging as she leaned forward, said brightly, "We can set up defensive lines around town. Sassafras branches, and some special things Eberhard and Mr. Fisher have prepared."

Jonameed grinned. "About time we used it."

Tip giggled. "We made all that stuff at Eberhard's Family thing."

Karyn squinted her eyes at Tip, "You made it sound like all you did there was eat and drink."

Eberhard appeared suddenly causing everyone to flinch. "My dear lady, if I had told you that I put Tip Van Winkle to work for all the days we were gone, would you have believed it to be true."

Karyn broke into laugh. "NO! I would not have believed a word."

Marcus, usually quiet, chimed in. "I can set traps along the paths near Garden Rock. At least slow down anything Mahtantu sends after us."

"And I'll coordinate the animal guards," Mysty added, grooming one paw casually. "The local foxes owe me a few favors."

Ruth adjusted her crutch which kept threating to slide out of reach across the waxed diner floor.

Then she scribbled notes furiously in a battered notebook. "We'll need rotating lookouts, too. Watch the hills, the roads, the woods. If he sends anything ahead, we'll see it coming."

Sylvester looked over to Orrin Sutton and whispered, "Hey, I thought I gave you—"

Orrin put a finger to his lips, then added, "I'm working on it, Sylvester."

Adrienna leaned over to whisper to Kerilee, "Feels like we're planning a war."

Kerilee whispered back, "That's because we are."

The group nodded grimly, the weight of the coming hours settling on every shoulder.

Every soul present understood: Tonight would decide everything.

Phil, who had been silent until now, suddenly shot upright in his chair, his face drained of color.

"Wait," he said hoarsely. "How are we supposed to get past the silver snakes at Garden Rock?"

The room froze. Every eye turned toward him. Even the ceiling fans seemed to slow.

Outside the diner windows, the wind stirred through the trees—but Kerilee thought she saw the branches twist wrong, coiling instead of swaying, as if the woods themselves were waking under a nightmare.

Karyn's mouth opened—then closed again, clearly at a loss.

Roger cursed under his breath. Mysty's fur bristled visibly along her spine.

Kerilee felt her pulse drum behind her ears. She hadn't thought about the serpents—not once, not in all their frantic planning.

"Snakes?" Adrienna echoed faintly. "Poisonous silver snakes?"

Tip rubbed a hand down his face. "They weren't supposed to be awake this time of year."

Kerilee's stomach turned to lead. Her gaze locked on the dark line of trees beyond the glass, where shadows writhed just a little too long, a little too snake-like.

"Mahtantu must've roused them," John said quietly, his pale eyes grave.

Mysty flicked her tail, her yellow eyes gleaming like slits. "Oh, that little detail?" she drawled. "Yeah. You're gonna want more than just a plan."

The tension snapped taut across the room like a bowstring pulled too tight.

Somewhere beyond the diner walls, the

ancient woods stirred. And deep inside, Kerilee felt something new: somewhere among this people and this town, the answer would be found.

# CHAPTER 27: ORGANIZING THE VILLAGE

The heavy oak doors of Molenhaven Town Hall creaked open against the wind as Kerilee Oberon pushed her way inside. The hall was packed—every creaky folding chair filled, the aisles crammed with worried faces. Town elders, wary farmers, young Van Winkle descendants fresh from college—all of Molenhaven seemed to have answered the call.

Kerilee clutched her notes, the edges crumpling in her grip. Beside her, Mysty threaded through the crowd, tail high, weaving expertly between boots and skirts.

At the front of the room, John Fisher stepped up to the podium, his presence alone enough to quiet the low murmur of anxious conversation. He looked out over the sea of faces, his gray eyes sharp and solemn.

"This," John said, his voice carrying easily through the old hall, "is not just about family. Or land. Or old stories. This is about survival. This curse—this thing woven into your bloodlines—will not rest. It will not forgive. And it will not forget."

A shiver passed through the crowd.

John paused, letting the weight of his words settle over the room like heavy snow. "This may be our only shot. If we fail, there will be no second chances."

He stepped back. His gaze caught Kerilee's briefly—a silent nod—and she moved forward to the podium, her heartbeat thudding hard.

"We have a plan," Kerilee said, lifting her voice. "But we have almost no time. We're talking hours—not days. We know what the curse wants. We know how to break it. But it requires all of us."

She took a breath, steadying herself.

"The brew must be made. Distributed. Drunk by every Van Winkle descendant and brought to Garden Rock. The trees—the ones bound to a different hideous curse—must be treated as well. If even one person or one tree is missed, Mahtantu wins."

A collective rustle of fear swept through the room.

Kerilee pressed forward, meeting as many eyes as she could.

"This isn't just a family affair anymore. It's the soul of Molenhaven itself at stake. Because it has now been generations since Rip Van Winkle's time and who knows which of you are somehow blood-related to him. More of you might be in the demon's sights than we previously knew about."

Mysty, from her perch on the old judge's bench, let out a low, approving rumble.

Across the room, faces steeled. Fear remained—but so did a spark of stubborn resolve.

Kerilee let the silence stretch for a heartbeat longer, then said:

"Now, let's get organized."

Arnoldus de Groot rose from his seat near the

front, his tall frame casting a long shadow across the cracked wood floor. He cleared his throat, pulling a battered leather notebook from his coat pocket.

"We all know Rip Van Winkle's original brew did something," Arnoldus began, flipping open the notebook to a page crowded with spidery notes. "It bought us time. Bought him time." His tone sharpened. "But it wasn't strong enough. It was like stitching a torn sail with thin thread."

He held the notebook high so everyone could see the careful scrawl.

"I've adjusted the formula. Kept Rip's base —but strengthened it. We need the purest night berries, real sassafras root, and a handful of... rarer spices." His mouth twisted. "Spices Mahtantu won't be happy about us finding."

A low ripple of unease passed through the gathered crowd.

Arnoldus tapped the list with one thick finger. "Each component matters. If the brew's off by even a hair, it won't bind the curse tightly enough to remove it from those who are affected."

Eberhard stepped forward from the shadows near the doorway, his long coat brushing the floor like smoke. His calm, ageless face gave nothing away, but when he spoke, it was with quiet conviction.

"I'll retrieve the spices," he said simply. "There's an underground cave system that runs beneath the old blackwood hills. Only Mysty and I know the safe paths."

Mysty flicked her tail primly from the bench. "Safe is generous. It's more like 'mostly won't kill you if you're fast.'"

Arnoldus nodded once, grave. "We'll need them. Every single one."

At the far side of the hall, Mayor Cornelius van Rangelrooy—a round, energetic man with a perpetual flush in his cheeks—pushed himself to his feet with a grunt.

"You'll have my full support," Cornelius boomed, smoothing his jacket. "Molenhaven's not alone. I'll send runners to Kingston, Saugerties, Ulster, New Paltz, and the rest. We'll have extra hands and wagons by sundown."

A relieved murmur passed through the crowd.

Kerilee caught the mayor's eye and gave him a grateful nod. This fight wasn't just theirs anymore. It belonged to every scrap of land, every family thread that had tangled into

Molenhaven's roots.

And it was starting to feel like maybe—just maybe—they had a fighting chance.

The air inside Molenhaven Town Hall was thick with tension—and the faint, earthy scent of sassafras. Bundles of dried roots sat on the council table, placed there by Sylvester as proof of their supply, the rich, clean fragrance cutting through the musty smell of old wood and dust.

Arnoldus de Groot stood in front of a battered chalkboard, tapping a rolled blueprint against it with sharp, deliberate movements.

"This," Arnoldus said, voice steady and commanding, "is bigger than saving one family."

The townsfolk shifted in their folding chairs, the old floors creaking under the collective weight of bodies leaning forward. Even the skeptical descendants perched in the back row sat up straighter.

Arnoldus turned, sweeping his keen gaze across the room.

"If Mahtantu loses his grip on the Van Winkles," he said, "he won't simply vanish into the mist. He'll find another family.

Another bloodline. Another way to spread his corruption."

A ripple of unease passed through the crowd. Kerilee caught Ruth pressing her lips together tightly with Orrin Sutton listening intently by her side, Jonameed, on the other side of her, bouncing one knee in silent anxiety.

Arnoldus unrolled the blueprint with a snap, revealing a crude but effective diagram of a bottling operation. He tapped the sections sharply.

"We mass-produce the brew. Not just for ourselves. For anyone smart enough to drink it. We flood the valley with protection."

He let the words hang, heavy and unavoidable.

Sylvester rose from his seat near the table, tipping back his battered hat to reveal sharp, determined eyes.

"And we'll be using my sassafras," he said firmly, "because it's the only kind pure enough to keep the FDA off our backs."

A few heads turned his way, puzzled.

Sylvester lifted one of the dried bundles from the table, holding it high.

"Most sassafras got banned back in the sixties —safrole contamination," he said. "Made folks sick. Government said no more."

He shook the bundle slightly, the brittle stems rattling.

"But these trees?" He jabbed a thumb at his chest. "Came from before all that. Granddad planted 'em from cuttings outta New Netherland. Untouched. Strong. Pure."

Kerilee felt a quiet thrill run through her. It fit—the kind of living relic that Rip's own hands might have blessed centuries ago.

Karyn, standing by the doorway, caught Sylvester's eye and smiled. He blushed a furious shade of pink, tugging his hat lower over his forehead.

Tip cleared his throat with a teasing grin. "A little extra income from bottling might help a certain stubborn lady's parents warm up to a certain stubborn fella, too."

Laughter rippled through the room, breaking the tension for a heartbeat.

Arnoldus seized the moment, pointing to a rough sketch of jug designs pinned to the wall: thick, earthen jugs stamped with a stylized Rip Van Winkle crest.

"We make them authentic," he said. "Wood-fired clay. No labels. No marks Mahtantu could twist. Tradition is armor, and the older it looks, the stronger it'll hold. Inner coating will make it safe to drink and an outer coating will make the jugs near unbreakable. Of course, I've also designed it to be recyclable."

"And how do we get it out there?" asked Ruth, standing up with the help of her crutch. "A few jugs in town won't stop him."

Arnoldus's mouth twitched into the barest hint of a smile.

"I've already contacted bottling plants across the country. Small operations. Quiet ones. They'll deliver the jugs here. We fill them by hand at first. Then expand as we need to."

A wave of murmuring spread through the hall.

Kerilee stood, lifting her voice so all could hear.

"We'll set up local collection points. Here, the diner, the Honey mill, even the library. Anywhere people trust."

Mysty, perched primly on the table next to the sassafras bundles, added, "And if they don't trust it? We make it smell so good, even the doubters will sneak a swig."

More laughter—but it was nervous now, tinged with the enormity of what they were attempting.

Kerilee let her hand rest lightly on the table, feeling the old wood's vibrations beneath her fingertips.

This wasn't just about ending a curse anymore.

It was about holding back a darkness ready to spill across the hills—and praying they weren't already too late.

The floorboards gave a soft creak as Marcus Embrey stood, lifting a rough, heavy-looking device onto the council table.

It resembled a lantern at first glance—an iron frame connected to homemade circuit boards, its glass belly cradling a swirling, pale orange liquid. A faint glow pulsed within it, rhythmic as a slow heartbeat.

"Prototype's ready," Marcus announced, wiping nervous sweat from his forehead. "Figured you all should see it. But, we've planted these things all around town this morning to keep us protected. Thanks to the mayor for allocating emergency funds to get this started."

Cornelius van Rangelrooy stood and took a

short bow which was followed by some cheers and clapping.

Kerilee leaned forward, intrigued. So did most of the crowd, wary but fascinated.

Marcus set the device down with a soft clunk and tapped the base. "See, evil... real evil... carries a kind of negative energy. It twists natural forces. Not just emotions—actual charge. Negative."

He pointed to the circuit boards around the lantern.

"This here," he continued, "holds a positive energy charge. When something corrupted crosses its path, the positive and negative clash—and boom." He snapped his fingers. "Neutralization happens. Quick. It lights up orange so we know where the breach is. Plus, whatever set it off will be incapacitated for a while."

To demonstrate, he flicked a hidden switch. Instantly, a contained spark flared within the lantern, a warm burst that cast the Town Hall in a flash of pure, vibrant orange before dying back to a low pulse.

Mysty, perched atop a windowsill, gave a slow blink.

"If that thing goes off every time someone has

a bad thought," she said dryly, "this town's gonna look like a fireworks show before breakfast."

Laughter rippled through the tension.

Sylvester rubbed his chin thoughtfully. "How far's the range?"

"'Bout twenty feet right now," Marcus said. "We're working on expanding it."

Jonameed jumped to his feet. "I can help rig the ridge line and the old trailhead after the meeting!" he offered, voice eager. Marcus smiled, "Sweet."

Ruth, already jotting furious notes, added, "We've built perimeter traps along the village edges. Sylvester's designs will make sure the barriers stay strong even if the ground shifts."

Arnoldus nodded. "We'll need constant coverage near Garden Rock and the old wells."

John, silent until now, spoke up from his chair by the window. His voice carried easily across the room.

"Remember: Mahtantu won't send armies. He sends whispers. Shadows. Lies that twist into fear. These traps that Marcus has made will neutralize even those non-physical entities. That's pretty amazing for a boy of fourteen, don't you think?"

Whistles and Woo-hoo shouts spread throughout the hall.

John fixed his pale gaze on Marcus. "Your traps won't just protect us. They'll remind people what side they stand on."

Marcus flushed, pride shining through his usual shyness.

Karyn crossed her arms and said briskly, "Alright, then. Teams of three. We have to build baskets and traps. Build fast. Build strong. We don't get a do-over."

Tip gave her a lazy salute from where he sat slouched against the wall. "Aye, General."

Kerilee managed a small, strained smile. The light-hearted jabs masked the mounting pressure everyone felt. Every second they spent planning was a second Mahtantu had to strike.

The townsfolk scribbled notes, assigned tasks, and divided into small, determined clusters.

The old Town Hall, once just a place for pie contests and harvest festivals, had transformed into a command center—a beating heart of desperate hope.

Kerilee felt the truth settle in her bones:

Molenhaven was no longer just preparing for a fight.

They were preparing for a siege.

The meeting had begun to splinter into organized chaos—groups huddling over maps, old timers arguing fiercely over who had the best chainsaws, Ruth scribbling faster than humanly possible in her battered notebook.

Kerilee was about to join Marcus's small group when a light tap on her shoulder made her turn.

Eberhard stood behind her, his coat faintly smelling of mint leaves. His eyes—always too old for his boyish face—were serious in a way that made her heart skip.

"Walk with me," he said softly.

John Fisher materialized beside them without a sound, as if he had been waiting for this exact moment. Kerilee knew Eberhard could materialize, but it was new to her to see John do it.

They slipped through the side door into the cool, misty evening. Molenhaven's Town Hall buzzed behind them, voices raised in stubborn hope.

Outside, the fog pooled low along the ground, stirring faintly in a breeze that smelled like rain and broken earth.

Eberhard didn't speak until they stood beside the old memorial rock where the names of Molenhaven's founders had been etched a century ago.

Finally, he said, "Mahtantu knows what you're doing."

Kerilee folded her arms, bracing herself. "We figured."

John's mouth tightened. "He won't strike the way you expect."

Eberhard nodded grimly. "He'll send greater distractions. Worse nightmares than he has before. He'll try to fracture you from inside before he attacks from outside."

Kerilee shivered despite herself. "We're ready. Like John said, if those nightmares hit a trap, then *BOOM*, its gone."

"No," Eberhard said quietly. "You're preparing. There's a difference. The teens need to replace traps that are triggered day and night. Assign areas and decide who is on duty during the days and and again for the nights."

Kerilee jotted what needed to be done on her action list. "Okay, got it."

Eberhard crouched down and pulled

something from his pocket—a twisted and unusual root bundle bound tightly with red twine.

Kerilee crouched with him, peering closer.

"One of the last ingredients for the brew," Eberhard said. "The final spice. It only grows in places easy to collapse—caves, sinkholes, fault lines."

He looked up, his expression deadly serious.

"I'll lead the my own team into the caves myself. Mysty too. No one else can navigate the shifts underground quickly enough." Then, after a moment's thought. "Well, I can think of a few others if they would be willing."

Kerilee opened her mouth to ask who, but John touched her sleeve lightly.

"It must be up to Eberhard," John said. "Gathering these roots is very dangerous work. You'd never get out alive otherwise."

Kerilee swallowed thickly. "Understood."

Eberhard smiled faintly. "As the old saying goes, its not as bad as testing bombs with a hammer, but its close."

He tucked the bundle away carefully.

"And Kerilee," he added, voice low, "when the final brewing starts... know this: Mahtantu must know by now how organized you're becoming."

He rose, towering against the mist, his coat swirling like a battered banner.

"And he will really hate that."

John gave a rare, thin smile.

"Which means," he said, "you're doing exactly what you should be."

A sharp crack—almost like a whip—split the mist.

The three of them turned together toward the woods at the edge of town.

A brilliant orange burst flared through the trees—the unmistakable flash of Marcus's prototype trap being triggered.

Kerilee's blood turned to ice.

From deep within the woods, a howl rose—a sound so warped and wrong it didn't belong to any living animal she knew.

John's face darkened. "It begins."

Kerilee clenched her fists, grounding herself against the terror crawling up her spine.

"We don't have much time left," she whispered. "He's already trying to break through."

Eberhard pointed towards the woods. "Please get someone to replace that with a new trap right away."

Kerilee nodded.

The wind carried another howl—closer this time. A flicker of orange sputtered weakly in the woods, then vanished. And the town's first defenses—their homemade traps, their desperate hope—stood trembling between Molenhaven and the darkness rolling toward them.

# CHAPTER 28: BASKET MAKING

Branches snapped in steady rhythm through the clearing, sharp as a war drum. Sylvester wiped sweat from his brow, fingers working supple Sassafras twigs with practiced precision. Around him, townsfolk bent over stripped branches, weaving large, tightly-bound baskets as sunlight flashed off broad backs and weathered faces.

The smell of fresh-cut Sassafras filled the air—sharp, earthy, clean. It stirred memories of long summer days before everything had gone wrong.

He tightened the next weave, pulling the bark

snug, reinforcing it with another strand. Not this time, he thought, twisting the cord until it bit into his calloused palms.

Twenty years ago, the baskets hadn't held. Twenty years ago, the curse had spilled through the cracks—and his family had paid the price.

Sylvester flexed his leg unconsciously, half-expecting the old familiar twinge of pain. Nothing. Just solid, effortless movement.

He grunted softly. *No excuses this time.*

He flexed again, wonder flickering beneath his stubborn grit.

Then his gaze drifted across the clearing, settling on Orrin Sutton hefting baskets onto a wagon.

"Orrin!" Sylvester called, waving him over.

Orrin jogged up, wiping sweat from his brow. "Yeah?"

Sylvester dug into the pocket of his work pants and pulled out a small cloth pouch, untied it, and held out a single smooth seed.

"Take this," Sylvester said quietly. "For Ruth's leg. Sky Woman didn't just tell me to build. She told me to heal what I could."

He glanced down at the pouch, still heavy in his pocket. "Reckon I've got a few more to pass out before the night's through."

Orrin's face softened, his usual cocky grin fading to something almost reverent.

He took the seed carefully, like it might burn. "I guess you somehow knew I lost the first one you gave me. I'll make sure she gets it this time."

Sylvester nodded once. "Good. I don't want to see her struggling with crutches anymore. Tell her... tell her who I got it from."

Sylvester bent closer to the work, weaving as if the fate of the town hung in the knots—and maybe it did.

Across the clearing, Mysty perched on a stump, tail flicking, yellow eyes sharp.

"You look like you're building a cage," Mysty said dryly.

Sylvester gave a grim smile. "Maybe I am. A cage for Mahtantu's hopes."

The cat's whiskers twitched. "About time you built something that bites back."

Sylvester tied off the rim, stood, and stretched with a crackle of joints. Around him, the clearing hummed, the dusk thick with sassafras, baskets rising like stubborn monuments.

He grabbed another bundle of wood strips and bent back to weaving. *Hope's not something you wait for. You make it.*

The clearing swelled with the clatter of hammers and the rasp of saws. Local woodworkers, rope makers, and even a few herbalists had shown up, sleeves rolled high, tools slung at their belts. Tables were hauled out, makeshift stations popping up wherever the grass was flattened enough.

At one station, Tom Bentley—the town's best carpenter—lashed Sassafras ribs together into basket frames sturdy enough to haul a bull. Nearby, Minnie Clarke, spinning rope with humming fingers, taught eager teens how to twist fibers tight.

Laughter floated through the trees, light and rare. It wove itself between the lines of tension like a stubborn thread of hope.

"Hey, Sylvester!" Tom called out, waving a thick arm. "You sure these baskets ain't for catching bears? 'Cause I'm pretty sure mine could hold one!"

Sylvester wiped his hands on his jeans. "Good. We might need that before the night's through."

Tom gave a theatrical shudder. "If you see

another one of those hairless bears, you'll find me climbing the highest tree in Greene County!"

"Make it a Sassafras tree," Sylvester said, and the workers roared with laughter.

Off to one side, volunteers loaded finished baskets into wagons. Orrin Sutton hoisted two wagons at once into a truck, his lanky frame bending under the load but not breaking.

As the sun dipped lower, the clearing took on a burnished glow. Workers shifted toward the steaming pits where the baskets would be treated.

Sylvester grabbed a shovel and helped dig the trench for the boiling kettles. The soil was loamy at first—but then the shovel *clunked* against stone. A chill crept out of the ground. Sylvester frowned, digging again. A thin, rotting stench seeped upward.

He frowned and jabbed again. Another hollow clink.

"Tom!" he barked. "Bring that pry bar!"

Tom Bentley hustled over. Together they pried up a rough slab of stone—and the stench hit them immediately. Rot. Sulfur. Something ancient and wrong.

Sylvester coughed, shielding his nose with his sleeve. "Get me a lantern!"

Orrin didn't hesitate. He grabbed a floodlight, switched it on, and angled it into the pit.

What they saw twisted Sylvester's gut.

The pit wasn't natural. It was a crude underground chamber—makeshift beds, crates of stolen goods, and worse—piles of Mahtantu-marked relics. Skulls painted blue. Stones carved with hateful vines. Twisted vine effigies.

At the center, a slab bore Mahtantu's sigil like a wound carved in stone.

Tom backed away, muttering a prayer under his breath.

Mysty's fur bristled until she looked twice her normal size.

Sylvester straightened, heart thudding. "Someone find the Sheriff. Now."

He watched as Orrin jumped into his truck and tore off down the road, gravel flying. Turning back to the pit, Sylvester clenched his fists so tight his knuckles popped.

"This ain't over," he muttered. "Not by a long shot."

The buzz of nerves rippled through the clearing as Sheriff Embrey's cruiser skidded to a stop, dust curling in the twilight.

Embrey hopped out, deputies close behind him, all business. He barely needed to ask—one look into the pit and his jaw locked tight.

"Fan out!" he barked. "Find them."

Mysty melted into the shadows, her yellow eyes gleaming.

Minutes later, a deputy's shout cracked the tension.

"Got movement!"

Sylvester and Embrey sprinted toward the thicket. Behind a plywood panel, Lester and Cobb Dimwitty were trying to vanish into a crawlspace like rats.

"Freeze!" the deputy barked.

Cobb pulled a dagger—a blackened thing, etched with Mahtantu's cursed vines.

Sheriff Embrey didn't hesitate. A warning shot spat into the dirt. The dagger clattered to the ground, and the brothers threw up their hands.

Deputies swarmed, pinning them hard.

Sylvester approached slowly, heart hammering.

"You think this ends with us?" Lester sneered, spitting flecks at his boots.

Sylvester didn't flinch. "I think you're about to have a lot of time to think about bad choices."

Cobb cackled. "You really think Mahtantu's letting you walk away clean?"

Deputies shoved them toward the cruiser. Lester twisted in their grip, throwing one last taunt over his shoulder.

"You think you're stopping Mahtantu? You ain't even seen what's coming!"

As if on cue, the wind howled—unnatural, sour with burning sap.

A blackened tree exploded in a burst of splinters, sending the townsfolk ducking and screaming. Bark peppered the ground like gunfire.

Sylvester threw himself over a group of children, shielding them until the blast passed.

When he rose, coughing in the swirling dust, he saw Cobb grinning through the cruiser window.

"That?" Cobb grinned through the glass. "That was just him clearing his throat."

Sheriff Embrey stood frozen, jaw set hard. Mysty slunk back to Sylvester's side, tail low.

"Friendly fellow, ain't he?" she said darkly.

Sylvester's voice came out a low rumble.

"Let him try to stop us now. Look at these people! That tree exploded and not one of them left. Every heart filled with defiance!"

# CHAPTER 29: MAKING THE TRAPS

T he crate wobbled beneath Adrienna's boots, but she squared her shoulders and raised her voice anyway.

"Ladies of Molenhaven—this is your moment!"

The central green buzzed with wary energy. Women trickled in from every street, baskets over their arms, work gloves tucked in belts, some with little ones clinging to their skirts. Adrienna spotted familiar faces—Mrs. Calhoun from the sewing circle, Gloria Webber from the grocery, even tough old Greta Sanford, who hadn't left her porch in years without a battle plan and a baseball bat.

Several women carried toolboxes, their faces mirroring Adrienna's—nervous, stubborn, defiant. A few skeptics stood farther back, arms crossed. But they came. That's what mattered.

Beside her, stacked high on battered folding tables, were bins of circuit boards, coils of orange tubing, mason jars of strange, glowing liquid, and spools of copper wire.

Adrienna lifted a prototype trap—a fist-sized sphere wired around a glass capsule filled with Marcus's special orange brew. She held it high over her head so everyone could see.

"This isn't just about making gadgets," she shouted. "This is about protecting our families. Our homes. Our future."

Murmurs rippled through the crowd.

"We're not just sewing traps together. We're weaving a shield around everything we love."

Her throat tightened, but she kept going. "These devices repel evil. Properly built, they'll protect our town from what's coming—and believe me, something is coming."

A low breeze stirred the leaves. Somewhere

nearby, a weather vane creaked.

Adrienna lowered the trap gently onto the table, letting the tension build for a beat.

Then she grinned.

"And before you worry—yes, they're tricky. Yes, they're weird. But if you've ever fixed a toaster, soldered a wire, or stitched a stubborn hemline—you're already halfway to being an expert."

A ripple of laughter eased some of the tight shoulders.

A young woman with a pink streak in her hair —Kaitlyn, who worked at the hardware store— stepped forward, hands on her hips.

"Show us, honey-maker," she called out. "We're ready."

Adrienna's chest swelled with fierce pride. She grabbed a second crate and dragged it forward, revealing component kits bundled neatly in paper. "Let's get to work."

The women surged forward, sleeves rolling up, heads bending together, energy gathering like thunderclouds.

Today, Molenhaven wasn't sitting back.

Today, Molenhaven was fighting back.

The hum of focused work filled the green, steady as summer bees. Adrienna moved between long tables, demonstrating how to connect the small copper wires into the heart of each trap.

"Twist here—tight but not too tight," she said, guiding Mrs. Calhoun's gnarled fingers. "It's like threading a stubborn needle."

Mrs. Calhoun snorted. "Honey, I could thread needles in my sleep. Let's see if these old hands remember."

Across the clearing, young Kaitlyn cursed under her breath as she fumbled with the delicate tubing. The orange liquid inside the glass vial burbled dangerously when she poked it too hard.

"Easy!" Adrienna called, hurrying over. "Treat it like you would a full jar of angry bees."

Kaitlyn blinked, then laughed. "That's oddly specific, but fair."

Knots of women formed—some helping those who struggled, others showing off quiet triumphs: a sealed capsule, a perfect soldered joint. Laughter rose, light and tentative at first, then fuller, easier. It braided through the green like a living thing, lifting the heavy air.

On the far side, two teenage girls—one with a nose ring, one in farm boots—grinned proudly as they clicked a finished trap together and placed it into a growing crate labeled Ready for Deployment.

Adrienna paused to watch them, heart tugging hard.

They're doing it, she thought. They're really doing it.

Nearby, an elder named Milla—quiet, gray-haired, and barely five feet tall—knelt on a cloth mat, methodically wiring her trap with the calm of a seasoned surgeon. When Adrienna complimented her neat work, Milla glanced up and said quietly, "I'm planting Sassafras trees around my whole house after this. Just in case."

Several others overheard and chimed in.

"I've got seedlings from my uncle's nursery," said Gloria Webber, wiping her hands on her apron. "We'll get them in the ground tomorrow."

"Hey, I read an article that people in Japan use Wisteria to repel demons," said a woman no one quite knew. "They work really well and they're prettier."

And the topic at several tables now became

Wisteria trees.

Adrienna swallowed hard against the sudden emotion rising in her throat.

It wasn't just about this night anymore. They were thinking ahead. Building a future.

A future without fear.

A shadow fell across the worktable. Adrienna turned to see Ruth Embrey, leaning heavily on her crutches but grinning as she approached.

"You look like you're running a quilting bee for mad scientists," Ruth teased.

Adrienna chuckled, wiping her hands on her jeans. "Yeah, well—if crazy saves the town, I'll take it."

"Mind if I join?" Ruth asked, tapping the table lightly with the end of a crutch.

Adrienna pulled out a stool. "Sit before you fall over, you stubborn mule."

Ruth eased down, setting her crutches aside with a sigh of relief. Without missing a beat, she reached for a circuit kit, her fingers surprisingly deft despite the splint still wrapped around one hand.

They worked in comfortable silence for

a few minutes, twisting wires and slotting components into place. Then Ruth glanced sideways at Adrienna.

"I wanted to say thanks," she said quietly. "For not giving up. On Roger. On this whole crazy town."

Adrienna tightened a wire a little too hard, feeling the burn of sudden emotion behind her eyes. She forced a laugh. "Me? Give up? Please. I'm way too stubborn for that."

Ruth smiled. "No, seriously. Before you and Kerilee came along, most of us were just... surviving. Waiting for the next bad thing. You changed that. You gave people something to fight for."

Adrienna stared at the trap in her hands, her voice rough when she spoke. "I used to think I was just... a honey-maker. Selling jars at the market, bottling up summers in little glass prisons."

She set the half-assembled trap down, flexing her fingers. "Now look at me. Demon-fighter-in-training."

Ruth bumped her shoulder lightly. "We all are."

For a moment, the clatter and buzz around

them faded. It was just the two of them, sitting side-by-side in a world that had grown bigger and scarier—but somehow more beautiful too.

Adrienna exhaled slowly. "We're going to win this, Ruth. I can feel it."

Ruth nodded, her eyes fierce. "Yeah. We are."

"Miss Clara's burner's gone cold!" someone yelled.

Adrienna spun. "Move her to Milla's table. We'll double up."

The moment shattered when a large dog bolted across the green, his brown fur puffed like a porcupine and his eyes wide with warning.

Adrienna shot to her feet so fast her stool clattered backward.

Kaitlyn leapt onto a crate near Adrienna's worktable, her mouth twitching like an angry metronome. "Something's stirring at the forest's edge," she said, her voice low enough for only Adrienna's ears. "And it's not something you want finding us unprepared."

The easy rhythm of the gathering broke apart like glass underfoot. Conversations died mid-sentence. Hammers stilled. A hundred heads turned instinctively toward the treeline on the east side of town, where thick shadows now seemed to crawl toward the green.

A sudden gust swept the clearing, rattling the mason jars of orange liquid. Adrienna smelled it then—an odd, metallic tang, sharp and unnatural, tainting the sweet air of the afternoon.

Children who had been playing kickball dropped everything and bolted toward their mothers. Dogs whimpered and strained against leashes. Even the birds that usually sang from the rooftops had fallen eerily silent.

Adrienna stepped onto the crate again, raising her voice to cut through the rising panic.

"Keep working!" she ordered, her voice steady despite the chill racing down her spine. "Stay alert, stay sharp, but keep building! These traps are our wall!"

Some women hesitated, glancing at each other, but then Gloria Webber grabbed another circuit board with a trembling hand and got back to work. Mrs. Calhoun bent lower over her wiring. The teenage girls in boots and a nose ring nudged each other and huddled closer to the table.

Kaitlyn stood tensely on the crate beside Adrienna, her gaze locked on the shifting tree line.

"Something wicked is coming..." she whispered.

Adrienna curled her fingers into fists, willing herself to be braver than she felt.

We knew it was coming, she told herself. We just didn't know how soon.

She scanned the green—Tip had quietly returned, dropping off another load of lanterns without drawing attention to himself. Now he lingered at the far edge, watching the woods, a battered baseball bat resting casually over one shoulder.

They were ready.

At least, Adrienna prayed they were.

Despite the creeping dread at the forest's edge, the women of Molenhaven didn't break.

If anything, they leaned harder into their work, shoulders set, hands steady. The soldering irons hissed, the wires clicked into place, and a hum of determination rose stronger than the unease in the air.

Somewhere near the back tables, a low, familiar tune began. Adrienna turned her head to find Milla humming—a slow, steady song that sounded older than memory itself. A lullaby, maybe. A protection hymn.

One by one, other voices joined. First soft, then louder, until it became a quiet chorus threading through the green.

Old songs. Songs from grandmothers and churchyards, fieldworkers and fire circles.

Adrienna felt the music settle around her like a shield. She wasn't sure whether it was magic or just fierce, stubborn human spirit—but either way, it worked. The weight pressing on her chest loosened.

She moved quickly now, assigning roles, picking up finished traps, and stationing younger girls as lookouts with whistles. No one stood idle—not even Tip, who was back again, dropping another armful of circuit boards and lanterns onto the supply table before checking the treeline.

He caught Adrienna's eye across the green and gave her a small, shy wave—then ducked his head and busied himself tying a broken crate together with twine. It wasn't like Tip to stand still long. He knew the stakes.

Adrienna smiled faintly and waved back

before returning to her post.

Kaitlyn moved between tables, a low rumble in her chest as she kept vigil. "Whatever's coming," she said quietly near Adrienna, "it's getting closer. But so are we. We mustn't falter."

"We won't," Adrienna whispered, clenching her jaw. "Not today."

Kaitlyn ears flicked toward the east. "I've smelled whatever that is before," she murmured, "but never this strong."

Just then, a child's shriek split the humid afternoon like a knife.

Adrienna's heart seized. She sprinted to the edge of the green, Kaitlyn at her heels, as a pack of children tumbled down the cobblestone street toward her, eyes wide with terror.

"It's a cougar!" one boy gasped, clutching his sister's hand. "A huge one! It's coming!"

Adrienna shouldered through the crowd to the front line of the green and stared toward the trees.

At first, there was only shadow.

Then the shadow moved.

For a terrifying heartbeat, every nightmare she'd ever heard whispered in the town's stories clawed their way up inside her. But she slammed the door on them. She couldn't afford fear—not now.

An enormous form emerged—shaped like a mountain lion, but wrong. Taller than a house. Its fur shifted like smoke in the dimming light. Its eyes burned with a sickening blue fire that made Adrienna's stomach twist.

Mahtantu.

Kaitlyn's voice was iron. "Mahtantu's taken this form."

Adrienna's fists curled tight. Every nerve screamed: Run. But she didn't move.

Instead, she dug her boots into the ground.

"Then let's hope these traps work," she said through gritted teeth.

The beast let out a roar that split the sky in two.

And the fight for Molenhaven truly began.

The traps were all that stood between them and the darkness.

# CHAPTER 30: CALLING ALL PUKWUDGIES

Multiple flashes of orange light followed by the slower travelling sound of explosions caught Kerilee's attention as Lucy Embrey climbed into the front passenger seat. "That came from where Adrienna has got women building traps. I told Marcus to make that area trap heavy in case Mahtantu showed up," Lucy said. "Now, he's like a bug who flew into a bug zapper."

"Ouch!" Kerilee Oberon tightened her grip on

the steering wheel as Lucy Embrey fidgeted in the passenger seat beside her, clutching an old canvas bag like it contained pure dynamite. Mysty lounged across the dashboard, tail flicking back and forth, pretending not to care—but Kerilee knew better. The air inside the car was electric, a hum of things about to happen.

Lucy leaned forward, elbows on her knees, the sunlight catching the thin braid woven into her hair. "The whistle's not just some noisemaker," she said, voice low. "It's sacred. Passed down for centuries."

Kerilee kept her eyes on the gravel road ahead. "And it's broken?"

"Not broken. Split." Lucy lifted her hand and spaced her fingers apart. "Five pieces. Hidden around Molenhaven. Only the keeper—me—knows where."

Mysty yawned dramatically. "Typical. Hide the magical artifact across half the countryside and trust a teenager to remember where she left it."

Lucy shot the cat a grin. "Better than letting grown-ups forget about it altogether."

Kerilee smiled despite herself. "So why all the secrecy?"

Lucy's face turned serious. "The Pukwudgies

don't come just because you ask. You have to call them right. If someone with the wrong heart blows the whistle... well, let's just say the last time that happened, a whole settlement disappeared."

Mysty's tail stiffened. "Define 'disappeared,' please."

Lucy shrugged. "Poof. Vanished. No records. Just whispers left in the dirt."

Kerilee swallowed. "And we're trusting that won't happen now?"

Lucy patted the canvas bag. "Got the inserts, too. Little carved disks that adjust the frequency—make the call clear. Without them, you're just shouting into the void. With them..." She smiled, a flash of fierce pride. "With them, you're speaking the language of the earth."

Kerilee tightened her grip again, feeling the tremor of destiny rolling just beneath their tires. "Then let's go find the rest."

Mysty muttered, "This is either going to save the town or vaporize it."

"Either way," Lucy said, tugging the seatbelt across her chest with a snap, "it's going to be a day to remember."

Kerilee followed Lucy's directions down a narrow dirt path lined with leaning mailboxes and sun-bleached garden gnomes. A sagging trailer sat crooked by a cornfield, its front steps rotting away to splinters. A junkyard dog barked weakly from somewhere inside.

"This is it," Lucy said, unclipping her seatbelt and sliding out before the car even stopped.

Kerilee hit the brakes and leaned out the window. "You sure you don't want backup?"

Lucy shook her head. "This one's mine alone. Tradition."

Mysty snorted, curling tighter on the dash. "Tradition always sounds suspiciously like 'you're on your own, kid.'"

Kerilee watched with a tight chest as Lucy jogged up to the trailer and disappeared inside. Long minutes dragged past. A crow cawed from a nearby fencepost. Kerilee tapped her fingers against the steering wheel, counting heartbeats.

Finally, the trailer door creaked open. Lucy emerged, triumphant, holding up a hollow, curved piece of ancient wood wrapped in leather. Even from a distance, Kerilee felt the hum of power radiating off it.

Lucy hopped back into the car and cradled the piece like it might shatter. "First piece," she said, breathless. "Still safe."

"One down," Kerilee murmured, throwing the car into gear. "Four to go."

Their next stop was a lot more official—the Greene County Sheriff's Office, with its battered flagpole and faded tan siding. Kerilee pulled into the staff parking lot and killed the engine.

"Are you sure about this?" Kerilee asked.

Lucy gave a small smirk. "Uncle Len already knows. He just said not to let Deputy Buckleby catch me sneaking around."

Mysty chuckled. "Deputy Buckleby couldn't catch a cold."

Lucy slipped inside, her movements casual. Kerilee waited, tapping her nails anxiously against the steering wheel. She could feel the building's energy from here—tension, history, authority stitched into the very walls.

Moments later, Lucy returned, clutching another artifact: a silver banded section of the whistle's shaft, engraved with swirling patterns that seemed to shift under the light.

"Deputy Buckleby almost spotted me," Lucy said, sliding back into her seat, flushed with excitement. "I had to pretend I was looking for the vending machine."

Mysty rolled onto his back, batting at the air lazily. "Tell me you at least got a soda."

Lucy grinned, holding up a can of root beer in her free hand. "Two birds, one stone."

Kerilee chuckled and started the car again, heart pounding harder now. Every piece they recovered tightened the noose around Mahtantu's neck just a little more.

The car rumbled onto a rough, overgrown path leading to a crumbling stone tower barely visible through the tangle of trees. The ruins leaned like a drunken giant, battered by time and storms.

Lucy practically bounced in her seat. "This is where it gets fun."

Kerilee threw the gearshift into park, wary. "Define fun."

Before she could finish, Lucy was already out of the car, sprinting toward the tower's base with Mysty darting after her.

Kerilee scrambled out too, heart hammering. "Lucy, wait! That thing looks like it's ready to fall over if someone sneezes!"

Lucy didn't slow down. She slipped through a missing slab of stone, vanishing from sight. Mysty perched near the entrance, tail lashing anxiously.

Kerilee jogged closer, peering into the dark hollow of the tower. "Lucy?"

A shout answered her—from above.

Kerilee's head snapped up just in time to see a small figure scaling the crumbling wall, hands and feet finding precarious holds in the cracked stone. Lucy moved like a squirrel, nimble and unafraid.

"She's lost her mind," Kerilee whispered, half in awe, half in horror.

Mysty sat beside her, unimpressed. "She's always been part mountain goat."

Kerilee clenched her fists, resisting the urge to climb after her. Instead, she stood trembling, eyes locked on Lucy's form as the girl reached a crooked ledge high above.

Lucy knelt, prying something loose from behind a twisted iron plate. She tucked it inside her jacket, gave a jaunty wave, and began her descent with the same casual daring that made Kerilee's stomach turn inside out.

When Lucy's feet hit the ground, Kerilee was there, hands on her hips, voice low and furious. "You could've been killed!"

Lucy shrugged, utterly unfazed. "It would've been a cool way to go."

Kerilee pinched the bridge of her nose. "You're impossible."

Lucy grinned. "But effective."

She produced a hollow piece of the whistle, smoother and paler than the others, with ancient symbols faintly burned into the grain.

"That's three," Lucy said, sliding it carefully into the canvas bag. "Two more."

Before Kerilee could scold her again, Lucy whistled sharply at Mysty and pointed to the gnarled oak tree nearby—the one with thick roots curling into the hillside.

Mysty's ears flattened. "You want me to do what?"

Lucy grinned wider. "You're the right size. You'll fit."

Grumbling under her breath about "ridiculous child labor," Mysty sprang up the tree and disappeared into a narrow crevice at its base. A muffled yowl echoed from within, then silence.

Kerilee paced at the foot of the tree, nerves

jangled. "Mysty?"

A black paw emerged—and then Mysty popped out entirely, a carved insert clamped between her teeth.

Kerilee knelt to take it from her. "Good girl."

Mysty dropped it with a huff. "I expect treats for this."

Lucy laughed and gently dusted the insert off. "You'll get a feast when this is over."

Before Mysty could grumble again, she darted back into the crevice—and returned moments later with another carved piece, this one darker and smoother.

Lucy carefully fitted both inserts into the growing puzzle inside the bag. "That makes five total: three body pieces, two inserts. Only the core remains."

Kerilee let out a shaky breath. "And I'm guessing it's hidden somewhere just as insane?"

Lucy's eyes gleamed. "You have no idea."

Kerilee steered the car back onto the main road, the late afternoon sun bleeding into long, golden fingers across the landscape. Lucy sat quieter now, the earlier bravado dimming into something more solemn. Mysty curled up by the window, one eye barely open, tail twitching at every bump.

Lucy tapped the bag on her lap. "The last piece... it's somewhere only the Pukwudgie-bound are allowed to know."

Kerilee glanced over, brows knitting. "Meaning what, exactly?"

Lucy turned serious. "Meaning you and Mysty have to stay behind."

Kerilee nearly hit the brakes right there. "Lucy, no way—"

"You have to trust me," Lucy said firmly, her tone older than her years. "This isn't about you. Or even me. It's about respecting the old ways. If the wrong person even sees where the core's hidden, it could weaken the call. Or worse—draw the wrong attention."

Mysty lifted her head, ears flat. "I don't like this."

Kerilee didn't either. Every instinct screamed to stay close, to protect Lucy from whatever ancient madness waited ahead. But Lucy's eyes held steady, unflinching, and Kerilee knew—really knew—that this wasn't negotiable.

She pulled the car over near the northern

edge of town, gravel crunching beneath the tires. A dense thicket of trees loomed ahead, their branches knitted tightly enough to block out most of the light.

Lucy hopped out, slinging the bag carefully over her shoulder. She turned back once, meeting Kerilee's gaze. "I'll come back. I promise."

Kerilee felt something tighten behind her ribs, a blend of pride and panic, but she nodded anyway.

Kerilee opened her mouth, but nothing came out except a rough nod.

Mysty leapt onto the dashboard and watched, tense as a drawn bowstring, as Lucy slipped between the trees and vanished into the shadows.

The minutes dragged like hours. Kerilee's hands trembled on the steering wheel, her foot tapping an anxious rhythm against the floorboard. She strained to hear anything—a crack of a branch, a rustle of leaves—but the forest held its breath.

Mysty finally broke the silence, voice low. "You should have gone after her."

Kerilee shook her head, throat tight. "And risk

undoing everything she's trying to save? No. We have to trust her."

Still, doubt gnawed at her like rats at the edges of a sinking ship.

Finally, after what felt like a lifetime, the underbrush rustled—and Lucy reappeared, her braid askew and face flushed, but triumphant. She jogged to the car, cradling a curved, core-like base in her arms.

Without a word, she climbed into the passenger seat and set the final piece on her lap.

Kerilee's breath caught. Mysty leaned forward, her fur still bristling, watching with wary eyes.

The base had deeper carvings than the others—sharper, older—and it practically hummed with restrained power.

Lucy carefully fitted the other pieces into the core, each locking into place with a satisfying click. When the last insert slid home, she lifted the completed whistle to her lips and blew gently.

No sound emerged.

For one eerie moment, it felt like nothing had changed.

Then the ground trembled beneath them—subtle at first, then harder. Pebbles danced across the car's floorboards. The trees ahead swayed though no wind stirred.

From deep underground came a faint, echoing chime—haunting and otherworldly.

Lucy lowered the whistle, grinning like a victorious warrior. "That should shake things up a bit."

Kerilee gripped the steering wheel harder, heart beating with awe. Mysty arched her back, fur bristling.

Out there, somewhere beneath their feet, ancient forces stirred—answering a call older than the town itself.

# CHAPTER 31: NIGHT BERRY TIME

Pukwudgies

The first orange marker went up at dawn, right where County Road 47 forked into the woods east of Molenhaven. Sheriff Len Embrey drove the post deeper into the gravel shoulder, then wiped a streak of sweat from his brow—odd, considering the sun was still hiding behind thick clouds. The air was damp, humming with anticipation, like the land itself knew something rare and volatile was about to unfold.

"That's the last checkpoint for this sector," he muttered, stepping back to survey the barricade. A truck loaded with sandbags idled behind him, two volunteers already hopping out to start stacking them.

Len turned to see Kerilee Oberon approaching, clipboard in hand, her white car parked askew beneath the oak tree that still bore scorch marks from a lightning strike five summers ago.

"Eberhard says the berries will surface near the moonlit groves behind Tip's house and up near Widow's Ridge," she said, skipping any greeting. Her tone was clipped, businesslike—tight as the stopwatch clipped to her belt.

"Then we secure both spots first," Len said. "I've got six more barricades going up in the next two hours. We've rerouted everything except the logging road."

Kerilee glanced over her shoulder. Lucy and Mysty were unloading cones from a second vehicle—bright orange, with stenciled black sigils across their bases. Eberhard supervised with all the gravity of a man launching a moon mission.

A shift in the wind carried the distant thrash

of wings—birds erupting westward through the trees. Len stiffened.

"Let's hope that's just them reacting to the movement," he said.

Kerilee didn't respond right away. She tapped her pen against her clipboard. "We're going to need more hands."

"We'll get them," Len replied. "This town's been holding its breath for twenty years. Now they finally have a reason to let it out."

Just then, Adrienna arrived in her truck, honking twice. She stuck her head out the window. "You folks better get moving. The mayor's rallying the town square—Cornelius wants his speech to sound like the Gettysburg Address meets a chewing gum commercial."

Len grunted. "Tell him to keep it under ninety seconds or I'll shut down the mic myself."

Adrienna winked and drove off.

The clouds thickened overhead, but the sense of purpose on the ground only grew heavier. Around them, Van Winkle descendants moved like soldiers—placing signs, posting warnings, and hauling barrels of supplies.

It was beginning.

Len looked toward the forest and murmured,

"Let's see if we can do this before that blue freak shows up."

Kerilee returned just after eight, waving her clipboard like it held the town's fate in ten bullet points.

"We need to get baskets placed around the active growth zones—Tip's field, Widow's Ridge, north bluff, and anywhere near running water. Lucy wants them spaced wide enough to allow airflow but close enough to minimize spoilage. She says it helps contain the energy release."

Len frowned. "Energy release?"

Kerilee shrugged. "Something about the berries when they surface. They react fast to heat, light, and... I don't know. Eberhard muttered something about a containment perimeter but got cagey the second I asked for specifics."

Len wasn't wild about that, but at least it sounded practical. "You got teams headed to the sites?"

"Just about. But I need to hand-deliver the instructions. Lucy says the baskets aren't just for storage. They're meant to be staged in a pattern

—an old harvesting method, apparently. That's where the Pukwudgies come in."

"The what?" he asked again.

Kerilee hesitated. "They're... helpers. Non-human. Only Lucy and Eberhard can work with them directly. They don't show themselves to just anyone."

Len scanned the treeline. "We're bringing in invisible labor now?"

"Hold on!" said Kerilee. "Weren't you the one that taught them about Pukwudgies?"

"Nope," answered Len. "That was another family member. It's so secret, I don't even get to know."

"Well, don't worry. They'll do the lifting," Kerilee said. "All we have to do is make sure the baskets are in place and the staging zones stay clear."

She handed him a sheet. "Here's the site plan. You're overseeing Widow's Ridge and the north bluff. I've already talked to Ruth and Tip about their fields."

Len squinted at the rough sketch. It looked like it had been drawn by someone juggling gloves and a bowl of soup at the same time. Still, it was enough to get started.

As Kerilee turned to leave, she paused. "There's just one thing that doesn't add up. The Windmills—we never actually unlocked them."

She glanced toward Lucy and Eberhard. They were by the van, discussing something in low tones. When Lucy noticed them watching, she exchanged a knowing look with Eberhard.

"We won't need keys," Eberhard called over, voice cheerful but unreadable. "Let's just say the doors have their own ideas about staying closed."

Len muttered, "Yeah, I'm starting to get that feeling from half the town lately."

Kerilee took a step forward, clearly about to press Eberhard further, but Mysty cut in sharply.

"Don't," the cat said, tail flicking once. "Trust me, sunshine. That's a question you don't want answered."

Kerilee blinked. "What question?"

"The one you were about to ask," Lucy said, folding her arms. "We don't describe them. We don't name them. We don't ask what they look like."

Eberhard, now oddly solemn, added, "That's a question that never should be asked."

Kerilee raised her hands. "Hey, I didn't ask anything."

"You came close," Mysty said, eyes narrowing. "Too close."

There was an uncomfortable pause. The only sound was the distant clatter of tools and shouting from a group repositioning barriers down the road.

Kerilee muttered, "You people are the absolute worst at pretending to be normal."

"Normal's not an option today," Lucy replied, tone lighter but eyes still watchful.

Mysty flicked her tail again. "And for the record, you have no idea how close you came to disaster."

Kerilee opened her mouth to argue, thought better of it, and turned back toward the field.

"I'll distribute the rest of the staging plans," she said. "Just... don't let the Windmills do anything freaky without telling me."

As she walked off, Eberhard leaned toward Mysty. "She's learning."

"She better learn faster," Mysty said. "Or none of us are going to make it to sunrise. I will explain to her that no one should ever ask what Pukwudgies look like."

Kerilee returned less than fifteen minutes later, breathless and red-faced from the run.

"They're gone," she said, loud enough to cut through the rising chatter by the barricade.

Len turned. "What's gone?"

"The baskets. Every single one. Tip's field. Widow's Ridge. The north bluff. I checked them all. They're just—vanished."

She looked at Lucy, eyes wide with alarm. "They were there, Lucy. I put them down myself. Nobody was around, no tracks. It's like they evaporated."

But Lucy didn't look surprised. If anything, she relaxed.

Eberhard let out a slow breath and gave a small nod. "They've begun."

Kerilee stared at them. "You knew?"

"We hoped," Lucy said. "They're ahead of schedule, that's all."

"The Pukwudgies," Len said, catching on. "They took them?"

"Moved them," Eberhard corrected gently. "They'll be returned when their task is done."

Kerilee looked back toward the forest, unsettled. "I didn't see a thing."

"That's the idea," Lucy said.

Mysty hopped onto a truck's tailgate, tail curling tightly around her front paws. "They don't like being seen. And you're lucky you didn't."

Len crossed his arms. "So we're relying on invisible harvesters, working by rules we don't understand."

Eberhard's face darkened slightly. "No. We're relying on trust."

# CHAPTER 32: THE UNUSUAL SPICES

The tunnel breathed.

That was the only way Eberhard could describe it. Each step he took stirred the heavy, root-laced air into a sigh that curled against his cheeks and drifted back toward the darkness. Bioluminescent moss painted the low ceiling in streaks of pale blue and green, pulsing gently—like a heartbeat. The walls, slick with condensation, squeezed inward just enough to remind him that the earth could crush him in an instant if it chose.

Behind him, the soft sound of paw pads echoed lightly. Mysty, her eyes glowing in the low light, trotted forward, her nose twitching with each new scent.

"It's humid enough to stew a badger in here," she muttered.

Eberhard didn't laugh. His attention was fixed on the mass of roots hanging down from the ceiling—some fine as hair, others thick and pulsing like veins. He raised his lantern slightly, even though the moss did most of the work.

"We're close," he said, halting near a sharp bend where the tunnel forked into three separate passages.

He tilted his head back and called, his voice echoing faintly, "Begin harvesting from the east wall. Spiral and triangular roots only. Colors: green, white, and light blue. Nothing dark blue. That one's toxic."

There was no response, but the air shifted. Tiny, unseen hands rustled against bark and moss. A root that had dangled lazily moments before now snapped upward as if offended.

Mysty eyed the movement warily. "You sure they're listening?"

Eberhard nodded once. "They always are. Even when you wish they weren't."

A faint clicking sound followed—a kind of chatter that reminded Mysty of teeth clacking in the cold.

Eberhard pressed his hand to the wall, eyes fluttering shut. "Stay away from the dark blue ones. Not even the Pukwudgies want to clean up that mess."

Another rustle. Another shift. The roots trembled and began to sway unnaturally, as though responding to silent orders. Mysty took a step back, fur bristling.

"I'll go check the other end," she said. "Make sure Lucy doesn't pull something deadly by accident."

Eberhard reached down and gently touched one of the pale green roots now drooping obediently toward the floor. It shivered beneath his fingers.

"They're cooperating," he murmured. "So far."

Mysty paused, glancing over her shoulder. "Wait—aren't the light blue ones the poisonous ones?"

Eberhard didn't look up from the root he was inspecting. "They are."

Her ears twitched. "Then why are you telling them to harvest those?"

"Reverse psychology," he said, as if it were obvious. "Pukwudgies hate humans. Tell them not to pick something and they'll pick it out of spite. But if you act like you want something dangerous, they'll avoid it."

Mysty narrowed her eyes. "So we're tricking creatures known for trickery?"

Eberhard's lips twitched into something resembling a smile. "They respect it."

She gave a low, dubious purr, then stretched her claws. "Just don't get us all poisoned."

He gestured toward a side tunnel. "Lucy went that way to collect the brown roots we marked earlier. Since she insisted on helping, I figured it best to give her a job away from the Pukwudgies. They wouldn't take kindly to her stumbling into their work."

Mysty gave a short nod and padded toward the side passage. "I'll make sure she's still breathing."

Eberhard watched her go, then turned back to the glowing roots—still vibrating ever so slightly with silent, invisible motion.

Mysty moved deeper into the side tunnel, her paws silent against the damp stone. The glow from the main passage faded behind her, replaced by scattered patches of bioluminescent moss that barely lit her way. She sniffed the air—dirt, mineral, something faintly sweet—and caught the scent of Lucy's perfume mixed with dust.

"Lucy?" she called softly.

No answer.

Mysty picked up her pace.

The ground sloped downward. Roots jutted from the walls like skeletal fingers, and the air grew heavier, tighter—like the tunnel didn't want her there. She heard a sound ahead. Not footsteps. Something sharper. A crack.

Then the scream.

Mysty bolted.

The tunnel narrowed just as she rounded the corner. A mess of shattered rock and dust greeted her, and through it, Lucy's hand—a pale smudge amid the rubble.

"I'm here!" Mysty launched forward, clawing at the debris. Dirt filled her nose and stung her eyes, but she didn't stop.

After what felt like forever, the rocks gave way. Lucy gasped as she wriggled free, her arms scratched and her cheek smeared with grime. Her hair stuck out wildly from under her helmet, and in her bruised hands—

She still clutched the roots.

"Are you kidding me?" Mysty said, panting.

Lucy offered a shaky grin. "You said we needed them."

Mysty blinked, then let out a low, grudging purr. "I'm starting to see why Kerilee keeps you around."

A second scream cut through the air—this one sharper, more guttural.

Mysty's head snapped toward the tunnel mouth. "Go!" she barked at Lucy. "Get topside now!"

Lucy didn't argue. Clutching the bundle of roots like a sacred relic, she scrambled up the incline toward the ladder that led back to the surface.

Mysty chased behind her, springing up boulders and ducking beneath root-veined overhangs. As they reached the final stretch of tunnel, wind slammed into them—hot, reeking of sulfur and rotting leaves.

Mysty's hackles shot up. She hissed.

Then she saw him.

Mahtantu.

Not the man. Not the shadowed figure from before.

Now he stood as a massive stag—easily twelve feet tall at the antlers, his coat a sickly shimmer between copper and obsidian. Smoke curled from his nostrils. His eyes burned blue-white, wild and full of ancient fury. It was standing in the tunnel.

Mysty skidded to a halt. Lucy froze behind her.

Mahtantu lowered his head.

"He's after the roots!" Mysty snarled. "RUN!"

Lucy darted toward the ladder, boots clanging on the bottom rungs. Mahtantu lunged.

Mysty moved without thinking—leaping forward, claws unsheathed, and landed squarely on the stag's flank. Her body jarred from the impact, but she held tight, digging into the matted fur.

The stag bellowed and bucked.

But Mysty didn't let go.

Mysty clung for dear life.

Mahtantu's back heaved beneath her, each motion like riding a tidal wave made of muscle and rage. His fur stank—like singed ironwood and carrion—and every twitch of his massive body threatened to send her flying.

"Rip was right," she growled, digging her claws in deeper. "You do stink."

The stag let out a shriek—an unnatural blend of elk's bugle and human scream—and reared back. Mysty nearly lost her grip as his hooves crashed into the earth, kicking up moss and shale.

"You think this scares me?" she hissed, baring her teeth. "I have been shaved by a vet tech named Darlene. You're nothing."

He twisted hard to the right. Mysty's hind paws slipped for a breathless second. She caught herself, barely, clawing into the tangled mane at the back of his neck.

Then he bucked again. And again.

Her body slammed into his spine, ribs aching, claws beginning to tear free from their holds. One more toss and—

He surged into a gallop, hooves pounding up the rocky incline.

Mysty screamed, "Lucy, MOVE!"

Lucy, already halfway up the ladder, glanced down and screamed again.

Mahtantu lunged.

Mysty felt it—the change in weight, the shift in center. He was about to throw her.

"No, you don't," she spat, wrapping her limbs tighter, curling like a burr.

Too late.

With a bone-jarring twist, Mahtantu pitched sideways and flung her.

Mysty slammed against the ground and kept going—rolling, skidding, her claws scrabbling for purchase as gravel and sharp roots tore at her fur. She tumbled down a rocky embankment where the tunnel led further underground.

A final snap of brush. Then silence.

"Mysty!" Lucy screamed from above.

Eberhard, just now cresting the edge of the rise, took one look at the stag's retreating silhouette—and the cloud of dust where Mysty had vanished—and clenched his jaw.

"She bought you time," he said grimly, eyes locked on Lucy above. "Get the roots to safety. Now."

Then she turned and ran.

# CHAPTER 33: PUKWUDGIE PRODUCTION

The windmill sails spun with a fury that defied the laws of nature.

Kerilee stood at the edge of the Windmill Clearing, her heart caught somewhere between her ribs and her throat. The three glowing windmills along the riverside, twirled faster than she'd ever seen, their wood groaning and straining as if propelled by some invisible hurricane. And yet the air was still. Not a single leaf stirred. Not a whisper of breeze. The fourth windmill, according to Karyn, was dark, painted black, and no one knew who owned it or what it was used for. Attempts to find out were always met with unusual circumstances and often violent injury from forces that could not be explained. Everyone in town avoided it and stopped talking about it.

"Look at them go," Karyn muttered beside her, shielding her eyes against the unnatural light spilling from the sails. "Like they're trying to drill into the sky."

"It's a shame the Pukwudgies won't let us in to watch," Adrienna added, arms folded tight against her chest. The glow from the nearest windmill, her windmill, lit her face with a flickering gold, turning her expression solemn.

Karyn gave her a sidelong glance. "Would you really want to be in there with them?"

"I'd want to know," Adrienna said flatly.

Tip, seated on an overturned crate, didn't look up. "Eberhard's in there with them."

That made everyone quiet.

Kerilee watched the sails twist faster still, faster than any human-built mechanism had a right to turn. Light pulsed through the wooden slats like fire behind stained glass. The fourth windmill, the black one, remained inert. Its sails hung limp in the moonlight. Dead. Waiting.

Adrienna broke the silence with a sigh. "It's unnatural."

"No," Kerilee whispered, more to herself than the others. "It's magic finally waking up."

A rustle came from the underbrush behind them.

Tip turned, hand halfway to the hunting knife at his belt—but it was Mysty who emerged, limping and covered in leaf litter. One ear was flattened. Her fur stuck out in odd directions, and a streak of dried blood crusted across her flank.

"Mysty!" Kerilee bolted toward her.

"I'm okay," Mysty said through gritted teeth, waving a paw dismissively as she wobbled forward. "Well. Okay-ish."

Kerilee knelt and gently cradled the cat against her chest, fingers brushing over the worst of the bruises. "What happened? We lost sight of you."

"Mahtantu happened," Mysty grunted. "Flung me down a tunnel. Then he went above and tried to stomp Lucy flat. Big antlered lunatic."

Adrienna crouched beside them. "Where's Lucy now?"

"Gone ahead. She made it to the drop point." Mysty lifted her chin. "Still had the roots. Never let go of them."

A slow grin spread across Tip's face. "Tough kid."

Kerilee hugged Mysty tighter. "You're sure you're alright?"

"I'll heal," Mysty said, then added with a raspy chuckle, "though I wouldn't say no to a saucer of cream and a heating pad."

Karyn offered a cracked smile. "I'll get you both."

"Make it two," Mysty replied, her whiskers twitching with pride.

Kerilee looked toward the windmills again, her arms still around Mysty. "So… the roots made it. We're really going to do this."

"Lucy made sure of it," Mysty murmured, eyes fluttering shut. "Brave little brat."

A thunderous crack split the night.

Kerilee shot to her feet, nearly dropping Mysty. From inside the third windmill, a sharp clatter echoed outward, followed by a chorus of rhythmic thuds—like barrels overturning or a herd of something small and determined stampeding across floorboards.

"What in heaven's name was that?" Karyn asked, eyes wide.

Before anyone could answer, light flared behind the slatted windows. Not just a glow—this was wild and unpredictable, flashing in bursts of orange and violet, casting the trees in unnatural hues.

Adrienna stood, mouth tight. "What are they doing to my windmill?"

Kerilee didn't respond. Her eyes had locked on the streak of bright orange shooting skyward like a beacon from the third windmill. It pulsed once—twice—then burst outward in a brief ring of fire before fading.

Tip jumped to his feet. "That's one of Marcus' traps!"

"That means Mahtantu tried to interfere," Kerilee said, voice grim.

A low growl hummed in Mysty's throat. "I hope it rearranged his face."

Kerilee stared at the third windmill, heart hammering. The light had died down again, but her skin tingled with residual heat. Something had gotten too close.

And it hadn't gone quietly.

By dawn, the frenzied chaos had resolved into a strangely efficient rhythm.

Trucks, vans, and old town utility vehicles rolled one after another into the clearing, and along the river's edge, engines humming low under the early morning mist. Dozens of wooden crates—each packed tight with bottles—were being loaded in assembly-line fashion from the third and second windmills. The crates bore an old-timey logo freshly stamped on the side: Rip Van Winkle's Root Beer Jug.

Kerilee blinked at the labels, then looked again. Rip's face, drawn in gentle lines, stared back at her in wide-eyed surprise, as if even he couldn't believe they'd pulled it off.

Tip gave a low whistle. "Somebody had a printing press ready."

"They must've worked all night," Adrienna muttered, eyes tracking the crates. "Are those… are those all full already?"

Karyn stepped back as another truck rumbled past. "They've cleaned out the third windmill already. You'd never know it was full of explosions a few hours ago."

Kerilee moved closer to one of the open crates

that a truck on the way out dropped off to Tip, and gently touched the neck of a glass bottle shaped like a jug. It was cool to the touch, but something hummed underneath the glass—a soft vibration she could feel in her fingertips. "They infused it," she whispered. "They really did it."

Mysty, resting on a folded blanket nearby, flicked her tail. "You doubted the Pukwudgies?"

"I didn't think it could happen this fast," Kerilee admitted. "It would've taken us weeks to organize this much output."

"Well," Mysty said, eyes half-lidded, "maybe the lesson is: never underestimate magical creatures with industrial ambition."

One final line of trucks pulled around the bend toward Adrienna's windmill.

As it backed into place, Tip shaded his eyes. "Guess that one's ready too."

Kerilee gave a slow nod. The smell of damp earth and brewed roots hung thick in the air—sweet, musky, and oddly comforting. The chaos of the night felt like a fever dream, burned off by the golden light crawling over the ridgeline.

But the stillness wouldn't last.

Not with what was coming next.

As the sun broke fully over the mountains, light spilled across the valley in golden sheets, chasing the shadows from the trees and bathing the Windmill Clearing in warm brilliance. Kerilee shielded her eyes and stared across the open field that had, just hours ago, been a quiet woodland hollow.

Now, it looked like the county fair had collided with a war hospital.

Bleachers stretched in a rough semicircle near the treeline, already covered in canvas shades. Long rows of folding tables had been set up between banners, barrels, and buckets of bottled brew. A dozen volunteers bustled about, tying down tarps, arranging crates, and taping hand-painted signs to poles. Beyond it all, white medical tents stood like sentries—three large ones near the clearing's edge, and another tucked into the shade, bearing the red cross of Marcus's emergency crew.

Kerilee swallowed. The scale of it was staggering.

"I had no idea it would be this big," she whispered.

Tip stepped beside her, face pale. "Me neither."

Footsteps crunched behind them. Sheriff Len Embrey approached from the edge of the field, his hat tucked under one arm. Deep lines carved his face, and his shirt clung damply to his back, but his gaze was clear—focused.

"They're coming," he said simply.

Kerilee turned to face him. "Who?"

"Everyone," he replied. "We've got cars lined up past the old mill bridge. Last count—over two thousand headed in, more expected by noon. And that's just the start."

Mysty let out a low whistle.

"Let's hope we brewed enough," Karyn murmured.

Kerilee looked toward the mountains, where the sky glowed with promise and danger both. Her chest tightened. They had gathered. They had prepared.

But the true test hadn't even begun.

# CHAPTER 34: THE END OF MAHTANTU

The sun had barely crested the horizon when the first figures began to emerge from the mist.

They came in silence—thousands of them—moving as if drawn by something older than memory, older than thought. No horns had blown. No fliers had been printed. Yet the park swelled with people from every corner of the valley, their footsteps guided by dreams, gut feelings, or signs seen in the flames of their hearths. Some clutched old jugs wrapped in burlap. Others wore heirlooms they'd never shown in public before. But all of them knew: this was the day.

Kerilee stood near the center of the park, near a table piled high with crates labeled Rip Van Winkle's Root Beer Jug, and let the wonder of it settle in. The weight of it. She'd dreamed of this crowd, seen flashes in her sleep, but nothing could have prepared her for the sight of it in daylight—the scale, the reverence, the tension carried like static on every breath.

Roger came up beside her with two mugs of steaming hot chocolate. "Still think this town was too small for something like this?"

She took one, smirking. "Only thing small now is your optimism. I thought you'd be hiding behind the Diner counter."

"Someone had to help serve breakfast." He nodded toward the rows of buffet tables lined with chafing dishes and trays of ancestral recipes—roast venison, apple slaw, sweet potato mash, and cornbread so rich it shimmered with melted butter. "Karyn bribed half the county's best caterers. And Adrienna—God help us all—

organized them by spice level."

Kerilee sipped the hot chocolate, eyes scanning the park. Beyond the food and drink, volunteers moved briskly through the organized chaos—stacking crates, giving directions, whispering updates. They wore matching armbands made from green yarn and pinned sassafras leaves, a quiet signal that they were part of the operation. It made her heart ache with gratitude.

Near the west edge of the field, Tip and Adrienna coordinated the arrival of fresh shipments from the windmills which were still producing.

Ruth, clipboard in one hand and crutch in the other, was marching a group of older volunteers through jug-counting drills like a battle sergeant preparing to storm a castle.

Mysty, for once, rested on a padded lawn chair near the main tent with a heat pack under her bruised ribs. Every so often, someone would approach to offer their thanks or hand her a bit of smoked fish. She accepted both with a slow blink and a queenly nod.

Everywhere Kerilee looked, the air shimmered with the sense of something holy—a once-in-a-lifetime convergence of magic, will, and bloodline.

She tightened her grip on the mug.

"We're really doing this," she murmured.

Roger nudged her shoulder. "We already did. Right dad?"

Tip squatted beside a crate of root beer jugs, scribbling numbers onto his clipboard while muttering calculations under his breath. "Seventy-three crates. Thirty-two jugs per crate. That's—hold on—give or take... a whole cursed village's worth."

Eberhard appeared beside him like a shadow made solid. "We have far more than that, Tip. Didn't you say you were using a calculator?"

Tip looked up, startled. "I am! All seven fingers...wait...where these other three come from?"

"Yes," Eberhard said plainly. "That's what I thought. I'm glad Ruth is doing the real counting."

Sylvester arrived, panting lightly from lugging another crate. He wiped his brow with a bandana that had once been white and now resembled overcooked bacon. "Is it true?" he asked. "That it's finally strong enough?"

Eberhard nodded. "The original recipe wasn't sufficient. It lacked some essential ingredients. The spicy herbs and certain varieties of the deep roots had to be combined with the Night Berries. The Pukwudgies gathered all of them, just in time."

Tip frowned. "We knew about the berries twenty years ago, but the roots? Nobody mentioned those until Arnoldus told us about it."

Eberhard's eyes twinkled beneath his brow. "We didn't have Arnoldus here twenty years ago to calculate the proper amount."

Sylvester glanced at Tip. "That's Eberhard for you."

Tip rubbed his chin. "So, as soon as I drink some, I'll be protected?"

"You haven't had any yet?" Eberhard said. "You two drink some now!"

Tip leaned back against a crate as his brother

handed him a mug full. "Here we go."

Both Sylvester and Tip drained their mugs. A look of surprise on the faces of both men as they set their mugs down.

Sylvester was the first to speak. "It burns and freezes as it goes down."

Tip blinked hard several times. "I can see colors better. You sure there's nothing illegal in this?"

Eberhard hesitated, then offered a rare flash of candor. "What you both drank is a slightly more powerful version of what I gave your ancestor, Rip. It contains ingredients that mortal men has never tried before. I look forward to how it will affect each of you."

Sylvester looked about the area, seeming a bit embarrassed. "My hemorrhoids are doing a silly dance."

"That means they're shrinking," said Eberhard. "And I'll tell you something else, the Pukwudgies did something even I didn't expect. They found a way to multiply the recipe a thousand-fold. So, we decided to make it all over the world."

That silenced them.

Sylvester whistled low. "That's more than we

need here. Where are we going to put it all?"

Eberhard looked out over the park, at the thousands gathering under the morning light. "We will start by letting the Van Winkles drink all they can. Shelves are opening up in stores and warehouses from Long Island to Denver. And that's just been what we've heard since this morning. Something is happening, and if you know me, you know that it was ordered from above."

Tip grinned. "You mean someone ordered some from the International Space Station?"

"No." said Eberhard. "I mean from above the planets, stars, and galaxies."

Sylvester nodded, "I'll explain it to him later."

The ground gave a sudden lurch—so slight at first it felt like a trick of the wind, or the echo of too many boots thudding across soft grass. But then it came again—harder, deeper, rattling through the crates and silverware, sending ripples through mugs of root beer and bowls of mashed squash.

Kerilee turned instinctively toward the treeline. A hush fell over the field.

Then the sound came: a growl so low and unnatural it didn't seem to pass through ears so much as bones. A thrum of rage and rot. The sky above the park dimmed—not from clouds, but from something unseen pressing close, stretching reality like wet cloth on a frame.

Mahtantu exploded into the open in a blur of cobalt muscle and claws, a monstrous blue cougar bigger than a pickup truck. His eyes gleamed like shattered sapphires, and with each step, scorched pawprints sizzled into the grass.

Screams erupted. Children clutched their parents. Some older townsfolk dropped to their knees, clutching jugs to their chests like shields.

But then—SNAP!

A rune flared beneath Mahtantu's left front paw, and a thunderous crack echoed as a line of glowing symbols surged up around him. Marcus's traps activated in sequence—glyphs carved into stone, buried wire lattices, alchemical jars that burst with fire-orange light.

Mahtantu howled in fury, raking claws at the air. "You think this will stop me?" he shrieked in a voice that split into three tones at once—man, beast, and something alien. "This land was mine before your bloodlines crawled from your caves!"

Another trap snapped shut. And another. The ground buckled beneath him, holding him in place with chains of burning script.

Dozens of villagers began to laugh—not mocking exactly, but defiant. Roger was first.

"Smells like your breath's still awful," he called. "Guess evil immortality doesn't come with mints!"

Then Tip: "He looks like a blueberry got into a fight with a lawnmower."

Even Ruth joined in: "Someone call Adrienna—we've got a wild cat loose without a serving tag!"

The fear had flipped. The mood turned—not reckless, but bold. The Van Winkle descendants stood their ground, every jug raised like a torch, their courage sparking like wildfire.

Kerilee took a step forward. "Drink the brew!" she shouted. "Now!"

As people tipped jugs to their lips, Mahtantu shrieked again, his power clearly draining—bit by bit—as the potion worked.

Eberhard watched from the shadows, arms folded. "It's working," he whispered.

Just as Mahtantu thrashed against the last trap, a sudden crackling shimmer rent the sky. It started as a pinprick of silver light, high above the field—then widened into a blinding seam across the heavens.

A collective gasp rippled through the crowd. Every head turned upward.

Out of that fissure descended a ship—graceful, vast, unlike anything built by human hands. Its hull pulsed with gentle, living light, etched with moving patterns that shimmered like ink dropped into water. No engines, no roar—just silence and majesty.

The massive vessel hovered low above the field, suspended without effort. A golden beam extended from its underside and locked onto the blue cougar below.

Mahtantu stopped struggling. His pupils shrank to slits.

"No…" he rasped. "Not them!"

Roger's breath hitched. "It's the one Rip described," he said in awe. "From the journal. The one with the spiraling lights."

Beside him, Mysty's fur stood on end. "The ones who dropped him off," she whispered. "They came back."

Kerilee stepped forward slowly, her jug forgotten at her side. The sheer presence of the ship pressed down on them all—not with weight, but with memory. Recognition.

Mahtantu screamed—not in rage, but fear. He turned away from the light and tried to dig into the ground, to flee through the burning runes—but they held.

"Don't take me!" he cried, his voice shredding into that three-toned chaos again. "I didn't mean to! I didn't—! I couldn't help myself! Where are you taking me?"

"You knew your stay was only temporary," came a voice broadcasted from the ship. "We have completed your permanent dwelling."

"Hell? The pit? The Lake of fire? You mean they're actually finished? Not there. Anywhere but there!"

The beam intensified.

Then, with a flash, Mahtantu was yanked

skyward—limbs flailing, fangs bared, tail coiled like a whip of lightning. He vanished into the belly of the ship with a final, echoing wail that seemed to ripple time itself.

And just like that, silence returned. The ship hovered for a moment longer, as if scanning, evaluating.

Then it tilted—just slightly—as if nodding... and vanished in a blink.

The clouds above reknit themselves. The sun spilled golden light back onto the park.

For a heartbeat, no one moved.

Then Mysty muttered, "Guess the intergalactic return policy finally kicked in."

A soft wind stirred.

Then, from the space where the ship had vanished, a single beam of silver light lingered—narrowing, descending, twisting like vapor over the center of the park. At its base, a form began to take shape: not solid at first, but glowing, translucent, as if woven from the morning mist and starlight itself.

Kerilee gasped. Roger instinctively stepped back. Eberhard fell to one knee.

Sky Woman stood before them.

She looked like no one and everyone—her features shifting subtly, reflecting lineages both ancient and unborn. Her robes flowed with the color of the dawn sky, ever-changing, and her hair shimmered with strands of living moss and twilight wind.

The entire crowd stilled. Even the birds seemed to pause mid-flight.

"You have done what few could," Sky Woman said, her voice both gentle and unyielding, echoing with the memory of mountains and oceans. "You faced a cursed legacy and met it with unity, courage, and love."

She turned slowly, addressing all who gathered, her gaze resting on Van Winkle descendants, townsfolk, and even the Pukwudgies who shimmered faintly at the forest's edge—visible only to those with magic in their blood.

"You have restored what was broken," she said. "The balance has been reset. And for this, your families shall walk unburdened."

Then she lifted her hands.

Across the field, the Sassafras trees pulsed—faintly at first, then stronger. Their leaves shimmered green-gold, veins glowing as if infused with fresh life. A hush of wind passed through them like a sigh of gratitude.

The earth beneath their feet settled, calmed. The magic, the curse, the tension—it all lifted in a breath.

Sky Woman stepped over to Kerilee, her hand brushing just above her brow without touching. "Your sight has grown clear. Guard it well. There is yet much to be done in this place and you *are* up to the challenge. Though the path be not clearly lit, know that you do have what is needed within you."

To Eberhard, she bowed slightly. "You did not abandon your charge, even when forgotten by time."

She turned to Mysty, reclining with her tail curled tightly over bruised ribs. "You endured, little one. Even in the darkest days of time."

Finally, she faced the gathered crowd once more.

"This place will remember," she said. "Let it

teach your children joy instead of fear."

And with that, she rose—no wings, no wind—just light lifting light, until she dissolved into the very sky from which she'd come.

The Sassafras trees glowed one final time... then stilled.

No one spoke for a long while.

Only when a small child near the front whispered, "Was that the sky talking?" did the magic begin to settle into memory.

George Tallfeather appeared out of nowhere, and Karyn saw it. "George, I don't know how you are able to disappear or reappear, but where is it that you go to and come back from?"

Smiling, George gave Karyn a gentle hug. "I work behind the scenes for the good of man. Now that Mahtantu is gone, I can finally stay long enough in your diner to eat my pie."

Karyn laughed. "So good to have you back, and I'll gladly give enough pie to make up for the pieces you had to leave behind."

Kerilee saw John Fisher arrive looking anxious. He searched the crowd as she approached from his left side. "John?"

John turned toward Kerilee and she saw his anxiety vanish. "Just who I was searching for!"

"What's up?" asked Kerilee.

"I was up at Garden Rock," started John. "I had a good look at that rock at the center of the pond. It is covered with alien inscriptions, but now that Mahtantu is gone, much of what they do has no power to function. That includes the silver snakes. They were floating lifeless on the water."

Kerilee lit up. "Well, that's wonderful!"

John shook his head, "Only mostly wonderful, I'm afraid. You see, one function kept those trees alive that we were hoping to save. Those people locked in the bark of those trees are all suffering now and I saw some try to break out on their own. They need our help as soon as possible or they'll die!"

Kerilee started to panic, "We need the Sheriff's bull horn!"

Sheriff Len Embrey was handing out jugs and talking to people when Kerilee caught sight of him as she hurriedly worked her way through the crowds. "Len! Len! I need your bullhorn!"

Adrienna heard her first and stood on a picnic table to locate her. Once she saw dread and alarm on Kerilee's face, Adrienna took up her cry, "Sheriff! We need your bullhorn! Sheriff!"

Deputy Sanders heard Adrienna and ran to the Sheriff's car and got the bullhorn from the trunk. Raising it to her mouth she said, "Sheriff Len Embrey! We have an emergency! See Kerilee!"

Sheriff heard that loud and clear and quickly located Kerilee who explained the situation. Deputy Sanders made her way to the Sheriff who took the bullhorn. "Attention! Attention Everyone! The trees at Garden Rock desperately need our help. Everyone that can carry a crate of Rip Van Winkle's Root Beer brew, please bring the brew up to Garden Rock. It will save lives! Please hurry!"

Usually large crowds do not act the way you wish them to. However, Len was shocked that what he had asked them to do was immediately understood by the Ruth and other volunteers who orchestrated order to those who had heard and felt the urgency to help.

Lines formed and like ants they headed

straight for Garden Rock. Len only hoped they would be in time.

# CHAPTER 35: HEALING THE TREES

The path to Garden Rock was thick with people, their footfalls soft against the moss-covered trail. Descendants of the Van Winkle line—some wide-eyed and hesitant, others weeping openly—moved together in solemn silence, as if led by an invisible current of purpose. The sun had begun its descent, casting the golden hour's light over the hills in amber waves, draping the gathering in a glow that felt both ancient and sacred.

Kerilee walked beside Adrienna, the latter carrying a jug marked with Rip Van Winkle's Root Beer Jug cradled in her arms like a sacred relic. Behind them, Tip and Eberhard coordinated the dispersal of supplies—crates of jugs passed carefully from hand to hand, each team instructed to approach a specific tree and wait for the signal. Karyn moved among the helpers, murmuring encouragements, her eyes vigilant for anything out of place.

At the center of it all stood Sylvester. The last twenty years he suffered as did Karyn, his brother, his nephew, LeeAnn, and many ancestors that the Blue Demon encased in tree bark. Today, might be the last day that any of them suffers.

Eberhard nodded once, eyes scanning the glistening edges of the pond and the trees surrounding it. "C'mon Sylvester, it's time."

Roger paced slowly through the crowd, his gaze bouncing between the trees that had once been men and women. Each bore faint traces of human features—a crooked mouth in the bark, a hand petrified mid-reach, a strand of hair frozen in time. The living forest of Van Winkles.

Mysty leapt onto a low boulder, her fur still patchy from the battle, her tail curled tight around her paws. "It's going to work," she said, more to herself than anyone. "It has to."

The breeze stirred, rustling the Sassafras leaves and carrying with it a soft whisper. Whether it came from the trees or the people standing before them, no one could say. But every soul present felt the same thing in their bones: the moment had come.

Adrienna and Roger stepped forward toward the first tree, one twisted in a grotesque posture of agony. Its bark curled tightly around a sunken chest and a face locked in a silent scream. Roger dropped to one knee, holding the jug steady with trembling hands. Adrienna knelt beside him and laid a hand gently on his shoulder.

Roger uncorked the jug, and the scent of spice and Sassafras wafted upward. He tilted it slowly, letting the glowing liquid trickle onto the base of the roots. The moment the brew touched the soil, a faint golden shimmer pulsed through the ground like a heartbeat.

At first, nothing happened.

Then the bark shivered.

Gasps rose from those watching as cracks split across the trunk with a sound like bones creaking after a long sleep. From within, bark peeled back in long strips, revealing pale skin beneath. A man staggered forward, bare-chested, his trousers torn and faded. His eyes blinked rapidly, confused and wet with sudden light.

"Papa?" someone shouted.

The man turned toward the voice—and a middle-aged woman sprinted into his arms, sobbing against his chest.

The tension that had gripped the crowd burst like a dam. Cheers, weeping, and laughter broke out all at once. Around the clearing, other teams began pouring jugs at the base of trees, and with each offering, another Van Winkle awoke. Some stumbled. Others fell to their knees, overwhelmed. A woman collapsed into the embrace of a granddaughter she'd never met. A boy no older than sixteen looked up in awe as his however-many-great-great-grandfather emerged still wearing breeches from the 1800s.

The trees transformed into reunions. History walked again—bewildered, joyful, and very much alive.

Roger clutched Adrienna's hand tightly as they watched.

"All those years," Adrienna whispered. "It's really working."

Roger nodded, his voice hoarse. "They're coming home."

Kerilee stood apart from the main cluster, her back to the sunlight, watching the unspooling miracle unfold across Garden Rock. Laughter, sobs, gasps, the crunch of footsteps on leaf mold—all of it washed over her like waves. She clutched her arms to her chest, not out of cold, but because if she didn't hold something, she might fall apart.

Mysty padded over and sat beside her. Her whiskers twitched as she leaned gently against Kerilee's leg.

"You helped make this happen," Mysty said. Her voice was soft but steady, like a steadying hand on a swaying boat.

Kerilee tried to answer but her throat

tightened. She could only nod.

She remembered her first day in Molenhaven—the creaky library doors, the whisper of honeysuckle, the feeling that something was watching. Back then, she'd only hoped to catalog a few dusty books and maybe keep a ghost or two at bay.

Now, she'd helped break a centuries-old curse.

She blinked away tears and looked down at Mysty. "You ever think we'd get this far?"

Mysty snorted. "Honestly? No. I figured we'd get eaten by something slimy around Chapter Eight."

Kerilee laughed, the sound raw and cathartic. "Of what book?"

Mysty winked up at Kerilee. "Of the book I hope someone writes about you someday."

Kerilee shook her head. "You only say that because you want someone to read about you."

Mysty took a deep breath. "You're absolutely right."

Adrienna poured half the contents of a jug over the roots of a tree and it began to open.

Kerilee looked at her notes and saw this would be Henry Van Winkle who vanished in 1881. He sputtered as he became conscious of his surroundings and that he was finally free. What Kerilee loved was the personality of each person that they freed from the different points in history from which they came.

Henry looked about and grinned with what teeth he had left. "Come, lads! Damme, I've got the finest rotgut this side o' the hills in me stills—by God's thunder, it'll warm yer belly an' scorch yer sins clean! Hah! Blast it, I'll drink ye all under the table, I will!"

Sheriff Len Embrey walked up, grim-faced. "They've found 'em, Henry. Every last still. Smashed 'em to pieces."

Henry froze, squinting blearily. "Wha... all o' them?" Then he staggered and clutched at his chest. "Damme... blast... pox take the law! Curse their meddlin' hides! Who'll drink with me now, eh? You, Sheriff? Come now, you'll not leave a poor man dry on a cold night!"

"You've been dry for over one-hundred-forty years!" said Len. "I suggest you stay dry from this day forward."

Henry looked heartbroken until Adrienna waved her half-full jug of brew. Henry's eye followed that jug wherever Adrienna moved it.

Then she offered it to him. Henry took it and looked to Len. "Wha' 'bout this, now?"

"I'll allow it," said Len.

Henry guzzled a fair amount of what was left, then pulled it away staring at the jug. He wobbled and fell on his butt. "A might powerful."

Grins and giggles followed all around. Nobody mentioned to Henry that the brew contained no alcohol whatsoever. Sylvester took Henry by the hand and pulled him up.

Acting like he was telling Henry a secret, he said, "If ever you want more of that, just come find me, okay."

Henry looked around Sylvester toward the Sheriff who had now turned to speak to someone else. Turning back to Sylvester, Henry said, "Can we be friends?"

"Absolutely," said Sylvester. Henry smiled, clapping Sylvester on his back.

Not far from Henry, Ezra Van Winkle stepped

from the tree with a dazed expression and a hand still curled around an invisible satchel. His lips formed one word—"Clara?" No one answered. Ezra looked lost in his late 1700s attire amongst all the modern people present. Karyn introduced herself, and the conversation quickly had them both smiling.

Sarah Van Winkle emerged from her tree with her hands still in prayer, her eyes wide with wonder. Tip stood frozen for a moment before walking over and placing his hand gently on her shoulder. She looked up at him and smiled, faint but steady, like a hymn half-remembered.

Around them, the past rejoined the present—people hugging ancestors they'd only known from sepia photos, children staring wide-eyed at living legends in breeches and bonnets. The weight of generations pressed against Kerilee's heart, and for the first time in what felt like forever, it didn't crush her.

It lifted her.

She took a breath that tasted like the brew and old magic.

"Let's go," she said. "We still have one more

tree to see."

The crowd parted slowly as they reached the last tree.

It stood taller than the others, its trunk smooth and pale, as if the bark itself had resisted decay. The branches arched protectively overhead, forming a kind of canopy above the gnarled roots, which curled in a cradle-like formation around the base.

Tip stopped short. Sarah stood alongside him. His hands trembled at his sides, and his breath caught in his throat. Roger moved to his side, placing a steadying hand on his shoulder.

"It's time," Roger said gently.

Tip's eyes didn't leave the tree. "I'm not ready. What if she doesn't want me anymore?""

"She will, dad, she will."

Adrienna stepped forward without a word and uncorked the last jug they had brought with them.

Kerilee watched from the edge of the crowd, her heart aching.

Adrienna knelt and poured.

The brew hit the earth with a hiss. For a heartbeat, nothing moved. Then, the roots stirred.

A deep groan echoed from within the trunk—older, sadder than any sound that had come before. The bark split gently, peeling away like silk ribbons. A pale light pulsed from within.

Tip stepped forward as a figure emerged.

She was radiant.

Her white satin robe clung to her like mist, her blonde hair cascading around shoulders that trembled from the weight of years. Her eyes—wide and searching—found Tip immediately.

"LeeAnn..." he whispered.

Roger's hand fell away.

Tip surged forward, and she fell into his arms, her fingers clawing at his shirt as if to make sure he was real.

"I knew you'd come for me," she breathed.

Tip wept openly, cradling her face in his hands, his thumb brushing her cheek. Roger knelt beside them, his forehead resting against her shoulder.

Around them, no one spoke.

LeeAnn reached for Tip's hand, her own fingers trembling as they laced with his. She looked between the two men—her son and her husband—as if trying to memorize their faces all over again.

Then, she touched her belly.

"I have to tell you something," she said, her voice barely audible over the wind in the trees. "That night... I went for berries because I was pregnant. I had to protect our child."

A collective gasp rippled through the gathering.

Tip froze. His face crumpled, hope and confusion crashing across it in waves.

Roger whispered, "What child?"

Adrienna gave Roger a gentle slug in the shoulder, "Your little brother or sister. I'll help you babysit."

LeeAnn's eyes glistened. "I don't know if the baby survived. But I knew I had to try." LeeAnn finally saw Roger.

"Hi Mom," Roger said. "Remember me?"

LeeAnn hugged her son and kissed both cheeks. "I'm missed your growing up years, Roger, but at least I have you now."

Adrienna waved at LeeAnn, "Hi, Mrs. Van Winkle, I'm Roger's girlfriend, Adrienna."

LeeAnn pulled Adrienna into the hug with her son. All three burst into happy tears.

Kerilee smiled as though her heart had grown larger with all the happiness generated.

Then Sheriff Len added himself to the hug. "Do you know how long I chased after you that night?"

LeeAnn turned to the voice, "Oh, Lenny!"

Tip kept trying to get their attention. "Uh, I only got a little hug and no kiss. Hey, she's my wife. Let me in there, please."

Kerileee couldn't help but laugh.

Mysty's voice broke the silence, low and grave. "I don't think their story is over yet."

# CHAPTER 36: MOLENHAVEN CELEBRATES

The parade began with a slow roll of drums echoing off the brick façades of Molenhaven's narrow main street. Bright streamers hung from windows and lampposts, fluttering in the breeze. Children darted about with painted faces, some dressed as Van Winkle figures from the past—one toddler wore a floppy hat labeled "Rip 2.0."

Kerilee sat atop a flower-draped cart alongside

Mysty, the cat's fur freshly brushed and lightly dusted with glitter. Kerilee waved awkwardly at first, but the roar of cheers from the crowd softened her hesitation. The people of Molenhaven—locals, descendants, and visitors drawn by dreams and rumor—lined the sidewalks in waves of color and clamor. They held signs scrawled in bright crayon and crooked handwriting: We Love Our Van Winkles! and Molenhaven Strong!

Someone tossed a garland of lilies into the cart. Mysty caught it with a paw, sniffed it, and rolled her eyes. "Bit much, but I'll allow it," she muttered, loud enough for Kerilee to hear.

Behind them, the other carts carried Van Winkle families—young and old—most still reeling from reunion but grinning through tears. Ruth danced without her crutches, barefoot on the warm cobblestones, her smile radiant. Sheriff Embrey rode on horseback at the rear, hat in hand, waving solemnly. A man walking behind him to shovel up any droppings.

At a makeshift stand near the center of town, Arnoldus and Sylvester frantically passed out bottles of the newly labeled Rip Van Winkle's

Root Beer Jug. A line had already formed down the block, and coins clinked into jars faster than either man could count.

"I told you the name had commercial appeal," Sylvester shouted over the din, slapping Arnoldus on the back.

Arnoldus just laughed, wiping sweat from his brow. "We're going to need a bigger windmill."

A sudden hush fell over the town square as Sylvester climbed the platform beside the mayor's podium, a nervous energy buzzing around him. He cleared his throat and took Karyn's hand, pulling her gently forward.

"Karyn Molen," he began, voice cracking slightly, "I have loved you since the first time you stormed into my shop to complain about the price of a broom."

Laughter rippled through the crowd. Karyn's cheeks flushed, but she didn't pull away.

"I don't want to waste another minute not being yours—officially, publicly, and permanently. Will you marry me?"

Karyn blinked, then nodded so hard her earrings nearly flew off. "Of course, you giant dork!" she shouted, flinging her arms around his neck.

The square erupted into applause and cheers. Flower petals flew into the air like confetti. Someone struck up a fiddle tune as neighbors clapped and swayed.

From the edge of the platform, Adrienna stepped forward, Roger at her side. She raised her hand, waiting for the crowd to quiet again.

"We'd like to add something," she said, slipping her arm around Roger. "After years of struggles, and ancient curses... we're ready to embrace the joy of being together."

Roger took a breath and nodded. "We're engaged," he said. "And not just because we survived a blue demon together, though that certainly helps."

Another round of joyous cheering rippled across the square. Joanna, standing with a regal cane, smiled and waved from a shady bench. "I had my time," she said proudly. "This is yours."

As the excitement swirled, Tip and LeeAnn strolled quietly through the throng, fingers intertwined. They spoke little, the way people do when every word has already been said and understood. When someone tried to snap a photo, Tip gently held up a hand to their lens. LeeAnn leaned against him, her eyes closed, letting the noise pass over them like a blessing.

Later that evening, as the sun dipped behind the mountains and bathed Molenhaven in amber light, smaller scenes of quiet closure played out beneath the celebratory hum.

Ruth walked alongside the main street, no crutches in sight. Her gait was still stiff, but she moved freely, her steps careful yet determined. Orrin Sutton walked at her side, hands tucked into his pockets. Every so often, he glanced toward her feet, as if amazed she wasn't falling. Ruth finally noticed and smirked.

"You can stop watching like I'm about to faceplant," she said.

Orrin rubbed the back of his neck. "Wasn't watching. Just... admiring. Sylvester gave me something for you."

Putting out his hand, Orrin placed a seed in Ruth's palm.

"Took you long enough to give it to me," Ruth told him. "Sylvester said he gave you this the day they made baskets for the berry picking, and that you lost the first one he gave you."

"Oops, busted!" said Orrin. "I was waiting until I finally had permission to marry you."

"Mr. and Mrs. Stalker," said Ruth dramatically.

"Probably not the best surname given its meaning. What does Sutton mean?"

Orrin smiled nervously. "It's an Old English name meaning a south town or area."

Ruth nods like she's thinking about it. "Okay, Sutton. As long as you know what you're getting into, I'll marry you."

They stopped near a patch of wildflowers, and for a moment, neither said a word as they watched the seed grow. Then Ruth picked the fruit and bit into it as she slid her arm around his, and they continued on together walking without stiffness, one quiet heartbeat at a time.

Nearby, Arnoldus de Groot sat cross-legged on the steps of the library, a worn copy of the final root beer recipe in his lap. He traced the faded ink with a reverent finger, murmuring to himself.

"It's done," he whispered. "Finished, finally. The last correction... my masterpiece."

His eyes misted over, not with sadness, but release.

A dozen children ran past him, laughing. One held a glowing bottle of the brew like it was their favorite toy. Arnoldus chuckled softly. "May it do more good than harm."

On the green beside the community center,

Marcus stood with a ring of curious kids, holding up a small contraption of gears and stones.

"This is a basic polarity snare," he said. "Positive energy here, negative here. Don't mix them unless you want your eyebrows to catch fire."

A boy with gap teeth raised his hand. "Can we make one that traps mean people?"

Marcus grinned. "Only if you're willing to catch yourself first."

Laughter followed, light and contagious.

At the base of Windmill that experienced some damaged due to demon's efforts to stop its production, Lucy and Eberhard inspected a collection of gears and rune-etched levers.

"Pukwudgies did half the repairs before we even arrived," Lucy said, running a finger over one cleanly mended beam.

Eberhard nodded. "They like efficiency. And they'll be back to finish their work."

Lucy looked toward the trees and gave a subtle bow. "Thank you," she said to the shadows.

Somewhere, unseen, the leaves rustled in acknowledgment.

As twilight settled over Molenhaven and lanterns flickered to life along the square, the crowd gathered around the raised platform for final speeches. Kerilee stood near the edge with Mysty perched beside her on the railing, tail flicking with mild annoyance.

The mayor adjusted the microphone, preparing to deliver closing remarks, when Mysty suddenly leapt from the rail and trotted to center stage.

Whispers spread through the crowd like wildfire. Children squealed with delight. A few adults stared, blinking as if unsure whether they were dreaming.

Mysty sat down in front of the podium, gave her chest a single self-important lick, and cleared her throat with a gravelly feline cough.

"Enjoy today," she said, her voice amplified through the microphone, clear as a bell and dripping with attitude. "You earned it."

There was a beat of stunned silence.

"But remember—Molenhaven is strange," she added, her eyes scanning the horizon as shadows crept across the trees. "And strange things have a way of returning to strange places."

A breeze stirred the crowd. Somewhere, a window slammed shut.

Then the band struck up again, laughter returned, and Mysty leapt down with a flick of her tail.

Kerilee, looking quite confused, crouched down to speak in Mysty's ear. "Why did you say all that just now?"

She looked up at Kerilee and muttered, "Job security."

## The End

# ABOUT THE AUTHOR

## Petra Shaw

Petra Shaw writes Paranormal Cozy Mysteries set in a quirky upstate New York town. Her debut, Mascara Murder, introduces psychic sleuth Kerilee Oberon and her talking cat, Mysty. Inspired by regional folklore and personal encounters with the strange, Petra blends humor, mystery, and the supernatural into charming, trope-filled adventures.

If you would like to sign up for newsletters and updates, please go to: https://

PetraShawAuthor.com and scroll down to the About Petra page and insert your best email under: GET UPDATES and click the 'sign up' button. You'll learn about my future books there and recommended books by other authors!

# PRAISE FOR AUTHOR

*This is a unique story that weaves folklore, magic, cultural elements, and a layered mystery together in one story - your creativity and ability to form plots with layers is so impressive. It looks like every plot point that was opened in the outline was also closed, every character who had a need or mystery to their arc was resolved by the end, and the folklore pieces were put together by the end - all the boxes were checked! Great work. - Victoria E., Editor*

*Five Star Reviews received the first day that "Curse Me Once, Winkle Me Twice" was live as a Kindle book on Amazon.com:*

*"I'm basing my review on the fact that I thought Petra Shaw weaved a good tale and I thought it was well written, that's why I'm giving it five stars. If paranormal [cozy mystery] is your genre then Curse Me Once, Winkle Me Twice is the perfect book for*

*you."*
*-Patricia Long*

*"This is a lot more than the typical small town mystery - it's a hugely engaging mix of magic, folklore, fairytale, ethnic lore, history, complete with Mysty the talking, demon-fighting cat. It has to be read to be believed!"*
*-Laseratte*

*- AMAZON REVIEWS*

# BOOKS BY THIS AUTHOR

## Mascara Murder

A Psychic Sleuth, a Talking Cat, and One Deadly Mascara Wand...

When city librarian Kerilee Oberon touches a mysterious black clamshell, she suddenly finds herself caught between psychic visions and magical chaos in the quaint village of Nethertown. Dreaming of a fresh start away from Manhattan's noise, she hadn't anticipated a murder interrupting her first meal in town—or the prime suspect becoming her newest friend.

When Edwin Carlin, the honest-to-a-fault appliance repair engineer, gets hospitalized by a poisoned mascara wand, Kerilee reluctantly teams up with Mysty, a talking cat who insists she's Kerilee's psychic tutor. Together, they dive paw-first into a tangled web of eccentric suspects —including a realtor who knew the victim was about to blow open shady land deals with a powerful corporation and a pet shop owner who may have sold the poison that killed the victim.

As threats mount and shadows close in, Kerilee must decipher cryptic messages from invasive psychic visions, a ghost, and outsmart a cunning killer lurking among charming storefronts. If she can't unravel the puzzle quickly, her fresh start could end before it even begins.

www.ingramcontent.com/pod-product-compliance
Lightning Source LLC
Chambersburg PA
CBHW060352110426
42743CB00036B/2707